Big Data Management

The big data paradigm presents a number of challenges for university curricula on big data or data science related topics. On the one hand, new research, tools and technologies are currently being developed to harness the increasingly large quantities of data being generated within our society. On the other, big data curricula at universities are still based on the computer science knowledge systems established in the 1960s and 70s. The gap between the theories and applications is becoming larger, as a result of which current education programs cannot meet the industry's demands for big data talents.

This series aims to refresh and complement the theory and knowledge framework for data management and analytics, reflect the latest research and applications in big data, and highlight key computational tools and techniques currently in development. Its goal is to publish a broad range of textbooks, research monographs, and edited volumes that will:

– Present a systematic and comprehensive knowledge structure for big data and data science research and education
– Supply lectures on big data and data science education with timely and practical reference materials to be used in courses
– Provide introductory and advanced instructional and reference material for students and professionals in computational science and big data
– Familiarize researchers with the latest discoveries and resources they need to advance the field
– Offer assistance to interdisciplinary researchers and practitioners seeking to learn more about big data

The scope of the series includes, but is not limited to, titles in the areas of database management, data mining, data analytics, search engines, data integration, NLP, knowledge graphs, information retrieval, social networks, etc. Other relevant topics will also be considered.

More information about this series at http://www.springer.com/series/15869

Yingxia Shao • Bin Cui • Lei Chen

Large-scale Graph Analysis: System, Algorithm and Optimization

 Springer

Yingxia Shao (iD)
School of Computer Science
Beijing University of Posts
and Telecommunications
Beijing, Beijing, China

Bin Cui
School of Electronics Engineering
and Computer Science
Peking University
Beijing, Beijing, China

Lei Chen
Department of Computer Science
and Engineering
Hong Kong University of Science
and Technology
Hong Kong, China

ISSN 2522-0179   ISSN 2522-0187   (electronic)
Big Data Management
ISBN 978-981-15-3930-5   ISBN 978-981-15-3928-2   (eBook)
https://doi.org/10.1007/978-981-15-3928-2

This Springer imprint is published by the registered company Springer Nature Singapore Pte Ltd.
The registered company address is: 152 Beach Road, #21-01/04 Gateway East, Singapore 189721, Singapore

Love the world as your own self; then you can truly care for all things.

– Lao Tzu

Preface

In this book, we will introduce readers to a methodology for scalable graph algorithm optimization in graph computing systems. Although the distributed graph computing system has been a standard platform for large graph analysis from 2010, it cannot efficiently handle advanced graph algorithms, which have complex computation patterns like dynamic and imbalance workload, huge amount of intermediate data, graph mutation, etc. Efficient and scalable large-scale graph analysis in general is a highly challenging research problem. We focus on the workload perspective and introduce a workload-aware cost model in the context of distributed graph computing systems. The cost model guides the development of high-performance graph algorithms. Furthermore, on the basis of the cost model, we subsequently present a system-level optimization resulting in a partition-aware graph-computing engine PAGE and present three efficient and scalable optimized graph algorithms— the subgraph enumeration, graph extraction, and cohesive subgraph detection.

This book offers a valuable reference guide for junior researchers, covering the latest advances in large-scale graph analysis, and for senior researchers, sharing state-of-the-art solutions of advanced graph algorithms. In addition, all the readers will find a workload-aware methodology for designing efficient large-scale graph algorithms.

Beijing, China   Yingxia Shao
Beijing, China   Bin Cui
Hong Kong, China   Lei Chen
December 2019

Acknowledgements

This work is supported by National Natural Science Foundation of China (Grant Nos. U1936104, 61702015, 61832001, and 61702016), National Key R&D Program of China (No. 2018YFB1004403), the Fundamental Research Funds for the Central Universities 2020RC25, Beijing Academy of Artificial Intelligence (BAAI), and PKUTencent Joint Research Lab. It is also partially supported by the Hong Kong RGC GRF Project 16214716, the National Science Foundation of China (NSFC) under Grant No. 61729201, Science and Technology Planning Project of Guangdong Province, China, No. 2015B010110006, Hong Kong ITC ITF grants ITS/044/18FX and ITS/470/18FX, Didi-HKUST joint research lab project, Microsoft Research Asia Collaborative Research Grant, and WeChat Research Grant.

The authors conducted the research at Beijing University of Posts and Telecommunications, Peking University, and Hong Kong University of Science and Technology. Since writing this book, we refer to the scientific researches of many scholars and experts. We would like to express to them our heartfelt thanks. We also received many graceful helps to shape the outcome of this book in one way or another from a number of friends. Hereby, we want to take this opportunity to convey our gratitude. We thank Prof. Junping Du from the Beijing University of Posts and Telecommunications, Prof. Junjie Yao from East China Normal University, and Prof. Hongzhi Yin from the University of Queensland for their countless discussions and comments. We also thank Prof. Yanyan Shen from Shanghai Jiao Tong University, Dr. Lele Yu from Tencent Inc., Dr. Xiaogang Shi from Tencent Inc., Dr. Ning Xu from Qingfeng Inc., Dr. Jiawei Jiang from Swiss Federal Institute of Technology Zurich, Dr. Zhipeng Zhang from Peking University, Dr. Xupeng Miao from Peking University, and the editors Celine Chang and Jane Li at Springer for their valuable feedback to improve the presentation of this book.

Last but not least, we would like to thank our own families for having our back. Thanks for all of your kind understandings and great supports.

Contents

Chapter 1
Introduction

Abstract With the rapid development of Internet of Things (IoT), mobile devices, and social networks, our world has become more connected than ever before, resulting in ubiquitous linked data, more generally, graphs. To discover the knowledge from the connective world, graph analysis is the de facto technique. In the consensus study report (National Research Council, Frontiers in massive data analysis. The National Academies Press, Washington, DC, 2013), National Research Council of US National Academies points out that graph analysis is one of the seven major computational methods of massive data analysis. A wide array of applications such as social network analysis, recommendations, semantic web, bioinformatics, intelligence surveillance, and image processing utilize graph analysis techniques to discover helpful insights. However, unlike decades ago, nowadays graphs are large, sparse, and highly dynamic. The classical methods of graph analysis become inefficient or even infeasible. Large-scale graph analysis becomes a problem that both industry and academia trying to solve. In this chapter, we first introduce the background of large-scale graph analysis and briefly review existing solutions; then we introduce three advanced graph analysis tasks which are popular and fundamental but not yet have efficient solutions on large graphs; third, we summarize the research issues of the large-scale graph analysis, especially for the advanced graph analysis tasks. Finally, we present an overview of this book.

1.1 Background

In the big data era, the graph size grows exceedingly fast. Graphs with billions of nodes and trillions of edges exist behind many Internet applications. For example, the Web graph indexed by Web search engines has more than 60 billion nodes, representing webpages, and one trillion edges, representing hyperlinks between pages [6]. The Facebook social graph has 1.39B billion nodes, representing Facebook users, and more than 400 billion edges, representing friendships between users [12]. By now there are also an array of popular graph data repositories, e.g., SNAP project [4], LAW [3], and KONECT [2], to share large graphs online for researches. Table 1.1 lists four graph datasets as representatives. Twitter2010 [5, 32]

© Springer Nature Singapore Pte Ltd. 2020
Y. Shao et al., *Large-scale Graph Analysis: System, Algorithm and Optimization*,
Big Data Management, https://doi.org/10.1007/978-981-15-3928-2_1

Table 1.1 Four examples of public billion-edge graphs

Dataset	Nodes	Edges	Graph type
Twitter2010	41,652,230	1,468,365,182	Social network
Friendster	65,608,366	1,806,067,135	Social network
UK200705	105,896,555	3,738,733,648	Web graph
PP-Miner	8,254,694	1,847,117,370	Protein–protein association networks

is a social network snapshot of Twitter site[1] in 2009, and it contains 41.7 million users and 1.47 billion social relations. UK200705[2] is a snapshot of web graph on the .uk domain around 2007-05. PP-Miner[3] is a collection of species-specific protein–protein association networks. Nodes represent proteins in different species (e.g., human, fruit fly, zebrafish, yeast) and edges represent direct (physical) protein–protein interactions, as well as indirect (functional) associations between proteins in a given species. Naturally it is an urgent task of developing new techniques to efficiently process and analyze these massive large-scale graphs.

Distributed computing is a standard solution for analyzing large-scale graphs. MapReduce [18] as a successful system for parallel data processing is adopted to handle large graphs. Cohen [16] first implemented some graph algorithms (e.g., triangle/rectangle enumeration, k-truss subgraph detection) on MapReduce. Then many other MapReduce-based graph algorithms are proposed for high scalability, such as connected component search [13], maximal clique detection [7], and general pattern graph enumeration [8]. But these solutions optimize each problem independently, and there are seldom common primitive operations. PEGA-SUS [28], a peta-scale graph mining library on top of MapReduce, formalized several graph algorithms as matrix–vector multiplication and introduced an important primitive, called GIM-V (generalized iterated matrix–vector multiplication). Similarly, GBASE [27] relied on matrix–vector multiplication to compute k-core, k-neighborhood, and induced subgraph. However, graph algorithms are often iterative and exhibit poor locality of memory access, very little work per vertex, and a changing degree of parallelism over the course of execution [37, 42]. MapReduce, a disk-based system, actually is not friendly for the graph analysis tasks, and the efficiency of the aforementioned solutions is always suboptimal.

Around 2010, researchers began to develop distributed systems for graph analysis, called *distributed graph computing systems*. Prominent examples include Pregel [38], Giraph [1], GraphLab [34], PowerGraph [20], GPS [46], Mizan [31], GraphX [21], and so on. These systems analyze the graph in-memory to mitigate the influences of poor locality of graph algorithms, and most of them parallelize the graph analysis tasks with bulk synchronous parallel (BSP) model [56]. They also

---

[1] https://twitter.com/.

[2] http://law.di.unimi.it/webdata/uk-2007-05/.

[3] http://snap.stanford.edu/biodata/datasets/10028/10028-PP-Miner.html.

provide graph-oriented programming primitives (e.g., vertex program and GAS vertex program) to transparentize the complex mechanism of distributed computing for users. With the help of graph computation abstraction, users can easily implement graph algorithms in distributed settings. In addition, distributed graph computing systems have fault-tolerant and high scalability compared to the traditional graph processing libraries, such as Parallel BGL [23] and CGMgraph [11].

Distributed graph computing systems indeed advance the techniques of large-scale graph analysis, and many internet companies (e.g., Google, Facebook, Alibaba, Tencent, etc.) have applied them to analyze large graphs. However, most of the systems only target simple graph algorithms, e.g., PageRank, breadth first search (BFS), single source shortest path (SSP), and cannot efficiently execute a lot of advanced graph algorithms which have complex computation patterns, especially for the matching-based graph algorithms. The root cause is that current graph computing systems cannot friendly handle the large amount of intermediate data and irregular workload distributions during the execution of the advanced graph algorithms, thus exhibiting low efficiency or infeasibility. In this book, we try to address the problem "how to achieve efficient solutions for the advanced graph algorithms on top of distributed graph computing systems." The basic idea to address the problem is that we develop novel system-level optimizations to balance the workload among systems, meanwhile, design algorithm-level optimization techniques to reduce the size of intermediate data. All the optimizations are introduced from the workload perspective. In the following sections of this chapter, we first introduce three advanced graph analysis tasks, then present the critical research issues to the above problem.

1.2 Graph Analysis Tasks

On the basis of existing literature about large graph analysis studies and experiments, there are tens of basic graph algorithms. The survey [24] roughly classifies them into eight classes, e.g., traversal, pattern matching, communities, etc. In this book, we aim to develop optimization techniques for advanced graph algorithms over distributed graph computing systems. We define the advanced graph algorithm as the one that has complex computation patterns, like huge amount of intermediate data, graph mutation, imbalanced workload, and so on. Therefore, we concentrate on algorithms that belong to communities and pattern matching and select three popular graph analysis tasks as representative of advanced graph algorithms. They are subgraph matching, subgraph extraction, and cohesive subgraph detection. Next we introduce the backgrounds of each task and present the characteristics of them.

1.2.1   Subgraph Matching and Enumeration

Graph matching problem is to determine whether there exists a bijective function between two graphs, which maps the nodes of one graph to the nodes of the other graph while preserving adjacency. It is also called *graph isomorphism* problem. Subgraph matching (or subgraph enumeration)[4] is a more general one, and it is an operation to find all the occurrences of one graph in the other graph. Concretely, given two graphs—a pattern G_p and a data graph G_d, subgraph matching is to enumerate all the isomorphism of G_p in G_d. Subgraph matching has been widely used in various applications, like the frequent subgraph mining, network processing, and motif discovering in bioinformatics [39]; information cascade pattern mining and triangle counting in social network [33].

On the basis of the aforementioned simple definition, there exist several variants of subgraph matching problems with different classification criteria. First, according to the usage of label information in graphs, we have labeled subgraph matching and unlabeled subgraph matching. The labeled one requires the mapped nodes and edges should have the same label, while the unlabeled one just focuses on the isomorphism of structure. One application of labeled subgraph matching is called homogeneous graph extraction [49], which uses a labeled graph pattern to extract new relation from the data graph, thus generating another graph for deep analysis. Second, according to different levels of restrictions of isomorphism, we have exact subgraph matching and inexact subgraph matching. The exact one finds the subgraph instances from a data graph that are exactly the same as the pattern graph. The inexact one aims to find approximate subgraph instances instead of the exact same ones, and it further has many variants. Here we introduce two of inexact subgraph matchings. The first one is proposed based on type-isomorphism by Berry [10]. All labels of nodes and edges are used, and type-isomorphic instances of a pattern graph are identified when labels match along some path in a data graph. The second one is called subgraph simulation [19, 26], which is defined as follows: Given pattern graph $G_p(V_p, E_p)$ and data graph $G_d(V_d, E_d)$, a binary relation $R \subseteq V_p \times V_d$ is said to be a match if (1) for each $(u_p, v_d) \in R$, u_p and v_d have the same label; and (2) for each edge $(u_p, u'_p) \in E_p$, there exists an edge $(v_d, v'_d) \in E_d$ such that $(u'_p, v'_d) \in R$.

Except for some inexact subgraph matching variants, most of the subgraph matching problem is NP-complete [17], resulting in expensive computation cost in centralized processing algorithms. Users turn to develop distributed approaches to handle large graphs. However, the following complex computation patterns hinder us to easily obtain an efficient solution.

- Large memory consumption. For subgraph matching problems, especially when all the matched subgraph instances need to be generated, the algorithm will maintain massive intermediate results (i.e., partial matched subgraph). In general,

---

[4]In this book, we use subgraph matching and subgraph enumeration interchangeably.

the size of intermediate results grows exponentially when the size of pattern graph increases. Therefore, when we run subgraph matching over a large-scale graph, we need massive memory to manage the intermediate size.

- Dynamic workload. In different matching iterations, the size of intermediate results, which is positively proportional to the workload of the iteration, varies, and we call this phenomenon dynamic workload. To achieve a better performance, we are required to design a mechanism to capture such dynamic workload.

1.2.2 Graph Extraction

Graph extraction is a basic operator that constructs a graph from the data source for subsequent analysis. Among various data sources, there is a type of data source called heterogeneous graph, which is also a graph. Considering that objects and their relations have diverse semantics in reality, heterogeneous graph is proposed to clearly represent the real-world multi-typed relations [54]. Analyzing heterogeneous graphs with these abundant semantics can lead to interpretable results and meaningful conclusions [54]. However, most of the existing classical graph algorithms, such as SimRank [51, 59], community detection [50, 52], centrality computation [44], etc., focus on homogeneous graphs. Directly executing them on heterogeneous graphs by simply ignoring the semantics of vertices and edges, the results will lose their values. Therefore, graph extraction is introduced as an important preprocessing step for analyzing the heterogeneous graphs, and it produces a subgraph which only contains single-typed edges and is called *edge homogeneous graph*.

In general, user creates the new single-typed relation by defining a pattern graph between a pair of nodes. In an edge homogeneous graph, there is an edge between two nodes when at least one subgraph between the two nodes in the heterogeneous graph is matched by the defined pattern graph. From this perspective, graph extraction is a special type of graph matching. Besides the single-typed relation, the extracted graph also requires quantitative attributes to represent the aggregated information between two nodes. For example, a simple attribute of an edge is the number of subgraphs matched by the pattern graph in the corresponding heterogeneous graph. Different aggregated functions result in different computation costs. This aggregation requirement causes the following new complex computation patterns compared with the pure graph matching task.

- The expensive computation of pair-wise aggregation. The existing graph matching solutions do not handle the aggregations. The aggregation will be expensive if we simply compute it by enumerating the matched subgraphs. This is because the number of matched subgraphs can be exponential as introduced in the above section.

1.2.3 Cohesive Subgraph Detection

Cohesive subgraph [57] is a kind of community in a large graph. It uses a measure of cohesion to find groups and is an important vehicle for the analysis of massive graphs. The most intuitive definition for a cohesive subgraph is a clique [36] in which each vertex is adjacent to every other vertex. However, for graphs in the real world, enormous cliques in small size exist while the number of larger cliques is limited, thus it is difficult to capture a meaningful subgraph for analysis, which reveals the clique definition is too strict to be helpful. Consequently, many other relaxed forms of cohesive subgraphs are proposed. n clique [9] loosens the distance constraint between any two vertices from 1 to n. n-clan [40] is an n-clique in which the diameter should be no greater than n, while n-club [40] is just a subgraph whose diameter is no greater than n without n-clique restriction. Furthermore, several other definitions weakening the degree constraint of the clique are also given. k-plex [48] is a subgraph containing n vertices in which each vertex connects to no fewer than $n - k$ vertices. In contrast, k-core [47] only requires that each vertex has at least k neighbors.

All the above cohesive subgraphs, except k-core, are faced with enormous enumeration problem and computational intractability. Taking the clique as an example, it requires exponential number of enumerations and will discover at most $3^{n/3}$ maximal cliques [41, 55]. Though k-core can be found in polynomial time and does not encounter the enumeration problem, its definition is too general and the results are not that cohesive [15, 60]. A most recently proposed cohesive subgraph is called k-truss [15] over social networks. In k-truss, each edge is involved in at least $k - 2$ triangles, and the number of involved triangles is called the support of that edge. k-truss is more rigorous than k-core but still looser than clique, and the results can be highly cohesive [15]. Especially in a social network, k-truss ensures that each pair of friends has at least $k - 2$ common friends, and this is consistent with sociological researches [22, 58]. Meanwhile, k-truss avoids the enumeration problem and can be detected in polynomial time as well. Given a graph, users are interested in finding all the k-truss in the graph by varying k, called k-truss detection. The following summarizes the characteristics of k-truss detection:

- Heterogeneous computation logic. Considering that the definition of truss is highly related to the triangle pattern, during the computation, we not only conduct the verification of truss, but also need to find/count triangles. Such different computation logic results in multi-phases graph computing, i.e., there exist multi-subtasks of graph computing.
- Graph mutation from the algorithm. When k varies during the detection, the edges involving computation varies as well. Carefully utilizing this mutation will boost the performance of the task.

1.3   The Research Issues

In this section, we discuss three research issues of designing an efficient graph algorithm over distributed graph computing systems.

Issue 1 *The distributed graph computing system is not a one-size-fit-all solution.*

Distributed graph computing systems indeed open up a new window for efficient large-scale graph analysis and enable many classical graph analytic applications (e.g., PageRank, graph traversal, single source shortest paths, etc.) to be run over real-world graphs. However, most of them are built through simply borrowing ideas of parallelization and are not carefully optimized for the features of graph analytic tasks. Actually, there are many factors (e.g., graph partitioning, data layout, execution engine, execution environment) affecting the performance of a distributed computing system. Taking graph partitioning as an example, we show that existing distributed graph computing systems, especially Pregel-like systems, are not friendly to the high-quality graph partitions. A high-quality graph partition implies that subgraphs are almost equal size and have seldom cross edges among them. It is a common sense that a high-quality graph partition does improve the performance of parallel systems by reducing the heavy communication cost [29]. However, based on our empirical studies, we found that Pregel-like systems are not aware of the quality of graph partitions, and cannot benefit from the high-quality graph partitions. In contrast, the systems suffer from the high-quality graph partition. Figure 1.1 shows the efficiency of PageRank on six different partition

Scheme	Random	LDG1	LDG2	LDG3	LDG4	METIS
Edge Cut Ratio	98.52%	82.88%	75.69%	66.37%	56.34%	3.48%

(a)

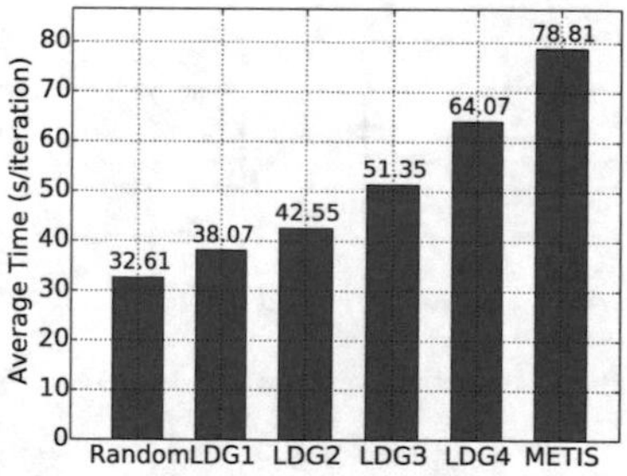

(b)

Fig. 1.1 The efficiency of PageRank in Giraph with different partitioning methods on UK200705. Random, LDG, and METIS are random hash partitioning, linear deterministic greedy graph partitioning [53], multi-level graph partitioning [29, 30], respectively. Edge cut ratio is the ratio of the edges crossing different subgraphs to the total edges of the graph. (a) Partition quality. UK200705 is partitioned into 60 subgraphs, and all the balance factors are less than 1%. (b) Computing performance

schemes over UK200705, and apparently the overall cost of PageRank per iteration increases with the quality improvement of different graph partitions. As an example, when the edge cut ratio is about 3.48% in METIS [30], the performance is about two times worse than that in simple random partition scheme where edge cut is 98.52%. It demonstrates that the performance of distributed graph systems is destroyed by the high-quality graph partition. As a consequence, we need to carefully consider the influence of graph partition schemes and other factors affecting the status of systems, develop novel architectures, and propose system-level optimizations to improve the performance across various situations.

Issue 2 *The diversity of graph structures in real-world graphs heavily influences the efficiency of graph algorithms.*

Compared to previous scientific computing graphs [25, 45] which are uniform-degree and regular, real-world graphs (e.g., Web graph, social networks) exhibit new characteristics, among which the most distinct one is power-law distribution [14]. The number of neighbors connected by a node is called the degree of the node. In real-world graphs, the degree distribution often satisfies the power-law distribution, i.e.,

$$p(d) \propto d^{-\gamma}, \tag{1.1}$$

where $p(d)$ is the probability of a node having a degree d and γ is a positive constant that implies the skewness of the degree distribution. A lower γ indicates that more nodes are high degree. Figure 1.2 visualizes the in- and out-degree distributions of Twitter2010 dataset. We can see that most of the nodes have low degree (<10), and there are few nodes which have degrees far larger than the average one. The distribution of other properties, e.g., triangle distribution [50], component size distribution [43], also has power-law property. Such skewed distribution

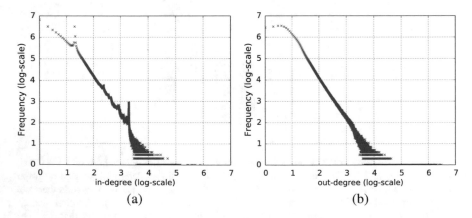

Fig. 1.2 The in-degree (**a**) and out-degree (**b**) distributions of Twitter2010 dataset. Both frequency and degree are in $log10$ scale

results in imbalanced workload and hinders classical graph algorithms to achieve a high performance for graph analytic applications. New graph algorithms must be designed for real-world large-scale graphs to obtain good parallel performance.

Issue 3 *The complex patterns of computation in graph analytic algorithms enhance the difficulty of parallelization.*

Graph algorithms like PageRank, BFS, SSSP, and Diameter Estimation [28] are extensively studied in the context of graph computing systems, and they are used as standard benchmarks to test the performance of the systems. These graph algorithms have simple computation patterns. Specifically, for most of them, each node in graph has constant or linear complexity of both the computation and communication, and in different iterations, the number of nodes involving computation is stable. We call them always-active algorithms [35]. However, real-world graph analysis also applies various other advanced graph algorithms, which bring in more complex computation patterns. According to the work [24], the graph algorithms can be classified into nine categories. Here we take subgraph matching as an example, which aims to find all the subgraphs in a large graph that are matched by a predefined graph pattern. The complexity of computation of a node not only depends on the large graph, but also depends on the graph pattern. High-degree nodes will generate massive matched subgraphs while low-degree nodes generate nothing. Overall, the computation workload is dynamic during different iterations and it transmits large volumes of intermediate results, which consumes huge amount of memories and communication bandwidth [52]. Besides the dynamic workload and explosion of intermediate results, other advanced algorithms [50] require graph mutation during the computation. These complex patterns of computation urge us to develop new framework for advanced graph algorithm-based analysis.

To sum up, existing systems are not a "one-size-fit-all" solution for distributed graph computing systems. They are not designed to efficiently support advanced graph algorithms processing large-scale real-world graphs. In this book, we will introduce our findings to optimize from both system perspective and algorithm perspective. The core idea of the optimization is to uncover the ingredients of the overall workload, build cost model, and design workload-aware techniques.

1.4 The Overview of the Book

In this book, we study the systems, algorithms, and optimization techniques for efficient large-scale graph analysis. As introduced before, we mainly focus on three issues existing in current graph computing systems, which are from system, data, and algorithm perspectives. On top of the detailed analysis of existing systems, we introduce a workload-aware cost model, especially for the Pregel-like graph computing systems. The cost model provides us helpful clues to develop high-performance systems and algorithms. The workload-aware cost model quantifies the cost from the workload source and the workload distribution. The workload source is the one that generates workload and affects the total amount of workload. It includes the number of iterations, the computation cost, and the communication cost. The workload distribution considers the factors affecting the balance of workload, such as graph partition schema, message distribution strategies, etc. Inspired by the workload, we propose both the system-level and algorithm-level optimization techniques which enable graph computing systems to process complex graph analysis algorithms efficiently. From the system perspective, we develop a partition-aware graph computing system. From the algorithm perspective, we focus on the aforementioned three graph analysis tasks and design three efficient and scalable graph algorithms. They are subgraph enumeration, graph extraction, and cohesive subgraph detection.

The whole book is organized into seven chapters. The brief introduction of each chapter and the relationships between them are described as below:

- Chapter 1: We introduce the backgrounds of large-scale graph analysis and describe the challenges and issues of existing graph computing systems. We also give a brief introduction of three popular graph analysis tasks, whose optimization techniques are introduced in this book.
- Chapter 2: We give a detailed introduction of distributed graph computing systems, mainly focusing on the system architecture, the computation models (e.g., vertex-centric, edge-centric, and subgraph-centric), and programming abstractions (e.g., vertex programming abstraction and gather–apply–scatter programming abstraction). Further, we study the cost of Pregel-like systems and propose the workload-aware cost model, which guides us to develop system-level and algorithm-level optimizations.
- Chapter 3: We present the system-level optimizations by introducing a partition-aware graph computing system, PAGE. The proposed system can smartly harness the quality of underlying graph partitions and achieve high performance of various graph analysis tasks by dynamically allocating computing resources (e.g., the degree of concurrency) on the basis of workloads.
- Chapter 4: We propose an efficient subgraph enumeration algorithm on top of Pregel-like graph computing systems. The basic idea of the proposed algorithm is to enumerate each subgraph in parallel, but it encounters the massive intermediate partial results which dominate the cost of the algorithm. We introduce several optimization techniques to reduce the intermediate result size and design an efficient and effective workload distribution strategy to balance the workload among workers.
- Chapter 5: We study the optimizations of parallel graph extraction over graph computing systems. In this work, we focus on a special graph extraction problem, where the input pattern is a line pattern. Since it consists of subgraph matching as its subtask, we optimize the intermediate result size via dynamic programming technique. For attribute computing, we apply partial aggregation techniques to reduce the amount of communications as well, thus guaranteeing the efficiency of the solution.

- Chapter 6: We focus on cohesive subgraph detection, especially k-truss detection, in this chapter. First, we introduce a subgraph-oriented computation model for the problem, since the problem can be an embarrassing parallel problem with overlapped graph partitions. Second, from the data perspective, we work out the *edge-support law* which ensures the efficiency of our solutions on power-law graphs.
- Chapter 7: We conclude our studies on the topic of large-scale graph analysis. Further, considering the development of this technique, we discuss two kinds of interested and important future works—large graph analysis on new hardware and large-scale graph neural network training.

References

1. *Giraph*, https://github.com/apache/giraph.
2. Konect: the Koblenz network collection. http://konect.uni-koblenz.de/. Accessed: 2019-11-22.
3. Law: Laboratory for web algorithmics. http://law.di.unimi.it/. Accessed: 2019-11-22.
4. Snap: Stanford network analysis platform. http://snap.stanford.edu/snap/index.html. Accessed: 2019-11-22.
5. What is twitter, a social network or a news media? https://an.kaist.ac.kr/traces/WWW2010.html. Accessed: 2019-11-22.
6. World wide web size. https://www.worldwidewebsize.com/. Accessed: 2019-12-05.
7. Ashraf Aboulnaga, Jingen Xiang, and Cong Guo. Scalable maximum clique computation using MapReduce. In *ICDE*, pages 74–85, 2013.
8. Foto N. Afrati, Dimitris Fotakis, and Jeffrey D. Ullman. Enumerating subgraph instances using map-reduce. In *ICDE*, pages 62–73, 2013.
9. Richard D. Alba. A graph-theoretic definition of a sociometric clique. Journal of Mathematical Sociology, pages 3–113, 1973.
10. J. W. Berry, B. Hendrickson, S. Kahan, and P. Konecny. Software and algorithms for graph queries on multithreaded architectures. In *2007 IEEE International Parallel and Distributed Processing Symposium*, pages 1–14, March 2007.
11. Albert Chan, Frank Dehne, and Ryan Taylor. CGMGRAPH/CGMLIB: Implementing and testing CGM graph algorithms on PC clusters and shared memory machines. In *Journal of HPCA*, pages 81–97, 2005.
12. Avery Ching, Sergey Edunov, Maja Kabiljo, Dionysios Logothetis, and Sambavi Muthukrishnan. One trillion edges: Graph processing at Facebook-scale. *Proc. VLDB Endow.*, 8(12):1804–1815, August 2015.
13. Laukik Chitnis, Anish Das Sarma, Ashwin Machanavajjhala, and Vibhor Rastogi. Finding connected components in map-reduce in logarithmic rounds. In *ICDE*, pages 50–61, 2013.
14. Aaron Clauset, Cosma Rohilla Shalizi, and M. E. J. Newman. Power-law distributions in empirical data. *SIAM Rev.*, 51(4):661–703, November 2009.
15. Jonathan Cohen. Trusses: Cohesive subgraphs for social network analysis. NSA., pages 1–29, 2008.
16. Jonathan Cohen. Graph twiddling in a MapReduce world. In *Computing in Science and Engg.*, July 2009.
17. S. A. Cook. The complexity of theorem proving procedures. In *Proceedings of the Third Annual ACM Symposium on the Theory of Computing*, pages 151–158, New York, 1971. ACM.
18. Jeffrey Dean and Sanjay Ghemawat. MapReduce: Simplified data processing on large clusters. In *Proceedings of the 6th Conference on Symposium on Operating Systems Design & Implementation - Volume 6*, OSDI'04, pages 10–10, 2004.

19. Wenfei Fan, Xin Wang, Yinghui Wu, and Dong Deng. Distributed graph simulation: Impossibility and possibility. *Proc. VLDB Endow.*, 7(12):1083–1094, 2014.
20. Joseph E. Gonzalez, Yucheng Low, Haijie Gu, Danny Bickson, and Carlos Guestrin. PowerGraph: distributed graph-parallel computation on natural graphs. In *OSDI*, 2012.
21. Joseph E. Gonzalez, Reynold S. Xin, Ankur Dave, Daniel Crankshaw, Michael J. Franklin, and Ion Stoica. GraphX: Graph processing in a distributed dataflow framework. In *Proceedings of the 11th USENIX Conference on Operating Systems Design and Implementation*, OSDI'14, pages 599–613, Berkeley, CA, USA, 2014. USENIX Association.
22. M. S. Granovetter. The Strength of Weak Ties. volume 78 of *Am. J. Sociol.*, 1973.
23. Douglas Gregor and Andrew Lumsdaine. The parallel BGL: A generic library for distributed graph computations. In *POOSC*, 2005.
24. Safiollah Heidari, Yogesh Simmhan, Rodrigo N. Calheiros, and Rajkumar Buyya. Scalable graph processing frameworks: A taxonomy and open challenges. *ACM Comput. Surv.*, 51(3):60:1–60:53, June 2018.
25. B. Hendrickson and J. W. Berry. Graph analysis with high-performance computing. *Computing in Science Engineering*, 10(2):14–19, March 2008.
26. M. R. Henzinger, T. A. Henzinger, and P. W. Kopke. Computing simulations on finite and infinite graphs. In *Proceedings of IEEE 36th Annual Foundations of Computer Science*, pages 453–462, Oct 1995.
27. U. Kang, Hanghang Tong, Jimeng Sun, Ching-Yung Lin, and Christos Faloutsos. GBASE: A scalable and general graph management system. In *KDD*, pages 1091–1099, 2011.
28. U. Kang, Charalampos E. Tsourakakis, and Christos Faloutsos. Pegasus: A peta-scale graph mining system implementation and observations. In *ICDM*, pages 229–238, 2009.
29. George Karypis and Vipin Kumar. Parallel multilevel graph partitioning. IPPS, 1996.
30. George Karypis and Vipin Kumar. Multilevel k-way partitioning scheme for irregular graphs. *Journal of Parallel and Distributed Computing*, 48(1):96–129, 1998.
31. Zuhair Khayyat, Karim Awara, Amani Alonazi, Hani Jamjoom, Dan Williams, and Panos Kalnis. Mizan: a system for dynamic load balancing in large-scale graph processing. In *EuroSys*, 2013.
32. Haewoon Kwak, Changhyun Lee, Hosung Park, and Sue Moon. What is Twitter, a social network or a news media? In *WWW '10: Proceedings of the 19th international conference on World wide web*, pages 591–600, 2010.
33. Jure Leskovec, Ajit Singh, and Jon Kleinberg. Patterns of influence in a recommendation network. PAKDD, pages 380–389, 2006.
34. Yucheng Low, Danny Bickson, Joseph Gonzalez, Carlos Guestrin, Aapo Kyrola, and Joseph M. Hellerstein. Distributed GraphLab: A framework for machine learning and data mining in the cloud. In *VLDB*, 2012.
35. Yi Lu, James Cheng, Da Yan, and Huanhuan Wu. Large-scale distributed graph computing systems: An experimental evaluation. *Proc. VLDB Endow.*, 8(3):281–292, November 2014.
36. R. Duncan Luce and Albert D. Perry. A method of matrix analysis of group structure. Psychometrika, 1949.
37. Andrew Lumsdaine, Douglas P. Gregor, Bruce Hendrickson, and Jonathan W. Berry. Challenges in parallel graph processing. *Parallel Processing Letters*, 17:5–20, 2007.
38. Grzegorz Malewicz, Matthew H. Austern, Aart J.C. Bik, James C. Dehnert, Ilan Horn, Naty Leiser, and Grzegorz Czajkowski. Pregel: A system for large-scale graph processing. In *SIGMOD*, 2010.
39. R. Milo, S. Shen-Orr, S. Itzkovitz, N. Kashtan, D. Chklovskii, and U. Alon. Network motifs: simple building blocks of complex networks. *Science (New York, N.Y.)*, 298(5594):824–827, October 2002.
40. Robert J. Mokken. Cliques, clubs and clans. volume 13 of *Qual. Quant.*, page 161–173, 1979.
41. J.W. Moon and L. Moser. On cliques in graphs. Israel J. Math., page 23–28, 1965.
42. Kameshwar Munagala and Abhiram Ranade. I/o-complexity of graph algorithms. In *Proceedings of the Tenth Annual ACM-SIAM Symposium on Discrete Algorithms*, SODA '99, pages 687–694, 1999.

43. M. E. J. Newman. Component sizes in networks with arbitrary degree distributions. *Phys. Rev. E*, 76:045101, Oct 2007.
44. Mark Newman. *Networks: An Introduction*. Oxford University Press, Inc., New York, NY, USA, 2010.
45. Alex Pothen. *Graph Partitioning Algorithms with Applications to Scientific Computing*, pages 323–368. Springer Netherlands, Dordrecht, 1997.
46. Semih Salihoglu and Jennifer Widom. GPS: A graph processing system. In *SSDBM*, 2013.
47. Stephen B. Seidman. Network structure and minimum degree. volume 5 of *Social Networks*, pages 269–287, 1983.
48. Stephen B. Seidman and Brian L. Foster. A graph-theoretic generalization of the clique concept. volume 6, pages 139–154, 1978.
49. Y. Shao, K. Lei, L. Chen, Z. Huang, B. Cui, Z. Liu, Y. Tong, and J. Xu. Fast parallel path concatenation for graph extraction. *IEEE Transactions on Knowledge and Data Engineering*, 29(10):2210–2222, Oct 2017.
50. Yingxia Shao, Lei Chen, and Bin Cui. Efficient cohesive subgraphs detection in parallel. In *Proc. of ACM SIGMOD Conference*, pages 613–624, 2014.
51. Yingxia Shao, Bin Cui, Lei Chen, Mingming Liu, and Xing Xie. An efficient similarity search framework for SimRank over large dynamic graphs. *Proc. VLDB Endow.*, 8(8):838–849, April 2015.
52. Yingxia Shao, Bin Cui, Lei Chen, Lin Ma, Junjie Yao, and Ning Xu. Parallel subgraph listing in a large-scale graph. In *Proc. of ACM SIGMOD Conference*, pages 625–636, 2014.
53. Isabelle Stanton and Gabriel Kliot. Streaming graph partitioning for large distributed graphs. In *Proc. of KDD*, pages 1222–1230, 2012.
54. Yizhou Sun and Jiawei Han. *Mining Heterogeneous Information Networks: Principles and Methodologies*. Morgan & Claypool Publishers, 2012.
55. Etsuji Tomita, Akira Tanaka, and Haruhisa Takahashi. The worst-case time complexity for generating all maximal cliques and computational experiments. Theor. Comput. Sci., pages 28–42, 2006.
56. Leslie G. Valiant. A bridging model for parallel computation. *Commun. ACM*, 33(8):103–111, August 1990.
57. Stanley Wasserman and Katherine Faust. Social network analysis: Methods and applications. Cambridge University Press, 1994.
58. Douglas R. White and Frank Harary. The cohesiveness of blocks in social networks: Node connectivity and conditional density. Sociol. Methodol., pages 1–79, 2001.
59. Zhipeng Zhang, Yingxia Shao, Bin Cui, and Ce Zhang. An experimental evaluation of SimRank-based similarity search algorithms. *Proc. VLDB Endow.*, 10(5):601–612, January 2017.
60. Feng Zhao and Anthony K. H. Tung. Large scale cohesive subgraphs discovery for social network visual analysis. PVLDB, pages 85–96, 2013.

Chapter 2
Graph Computing Systems
for Large-Scale Graph Analysis

Abstract Since Google introduced the first distributed graph computing system Pregel, many similar systems are proposed. The distributed graph computing systems become a standard platform for large-scale graph analysis. Compared to the previous graph processing libraries, the new systems have the advantages of scalability, usability, and flexibility. In this chapter, we briefly review the basic concepts of the distributed graph computing systems, including the architecture, execution flow, and programming abstraction and computation models (e.g., vertex-centric, edge-centric, subgraph-centric, etc.). In this book, we concentrate on the vertex-centric computation model and then describe two excellent and popular programming abstractions—vertex programming abstraction and gather–apply–scatter (GAS) programming abstraction. Finally, we introduce the workload-aware cost model which classifies the factors influencing the performance into two types—workload source and workload distribution. The model helps to estimate the workload for a distributed graph computing system and guides us to optimize the systems and algorithms smartly.

2.1 Distributed Graph Computing Systems

Distributed graph computing is a prominent framework for large-scale graph analysis. Before the big data era, actually many parallel graph algorithm libraries have been proposed, like Parallel BGL [8] and CGMgraph [4], which are designed for high-performance computing and built on top of Message Passing Interface (MPI) [21]. Although they provide an approach to handle scientific computation graphs efficiently, they are lack of friendly graph programming interfaces and fault tolerance, which hinder them from being extensively applied in large-scale environment. Another type of parallel graph processing libraries like SNAP [3] and Green-Marl [10] is restricted to shared memory and cannot be applied onto commodity clusters and clouds. In 2010, Google first introduced a novel distributed graph computing system Pregel [11]. Pregel can not only apply to large-scale clusters, but also hide the complex distributed logic easing users to implement a parallel graph algorithm. Later on, many similar systems have been designed.

© Springer Nature Singapore Pte Ltd. 2020
Y. Shao et al., *Large-scale Graph Analysis: System, Algorithm and Optimization*,
Big Data Management, https://doi.org/10.1007/978-981-15-3928-2_2

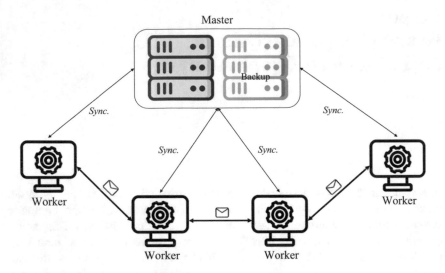

Fig. 2.1 The master–worker architecture

These systems focus on scalability and usability and provide flexible graph-parallel abstractions.

As a kind of distributed system, most of the distributed graph computing systems follow the master–worker architecture [1]. As shown in Fig. 2.1, the master is responsible for aggregating global statistics and coordinating global synchronization. In general, the communication between worker and master is low. The workers are responsible for computation and can communicate with each other. The data graph to be analyzed is partitioned and distributively stored among workers' memory, thus parallelizing the workload. In addition, to provide fault tolerance, the master is usually equipped with a backup server.

Execution Flow In the distributed graph computing engine, graph algorithms are executed based on bulk synchronous parallel (BSP) model [20]. That is, a graph algorithm is decomposed into a series of supersteps (or iterations); between successive supersteps, a synchronization is introduced to ensure that all the computation and communication are finished in the current superstep. Figure 2.2 illustrates the supersteps in the BSP model.

In a superstep, vertices in the graph execute user-defined functions with the states from their in-coming neighbors in parallel and send the latest state to their out-coming neighbors through message passing [21]. When there is no new message in the system, the whole graph algorithm finishes. Considering that there are various graph algorithms, in which not all vertices are involved for the computation in every superstep, therefore, during the computation, the vertices in the graph actually have two states: active and inactive. All the active vertices participate in the computation, while the inactive ones become active when they receive messages. More specifically, in each superstep, a vertex finishes its computation and there is

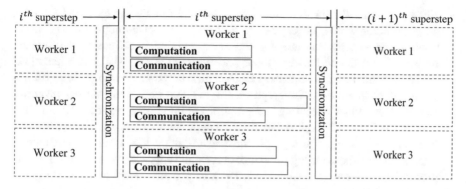

Fig. 2.2 Supersteps in the BSP model

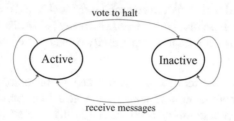

Fig. 2.3 The state machine of a vertex in graph computing systems

no more subsequent computation, it will vote to halt and set itself to be inactive. In the subsequent superstep, the system will not execute the algorithm logic of inactive vertices unless it finds that the inactive vertices receive messages from the previous superstep. Figure 2.3 visualizes the state machine of a vertex between active and inactive states.

Programming Abstractions and Computation Models To release users from the above tedious communication protocol, distributed graph computing systems provide flexible graph-oriented programming abstractions. With the help of the abstractions, users only need to define the computing logic (i.e., user-defined function) for local graph view (e.g., vertices, subgraphs) and the systems can automatically parallelize the algorithm. For a distributed graph computing system, the underlying data layout and computation model determine the programming abstractions. So far, many different computation models have been proposed. The survey [9] presents a relative complete list. We discuss several classical computation models below.

- *Vertex-centric model*. Vertex-centric model is the most mature distributed graph programming abstraction and most of the distributed graph computing systems have been implemented using this model [5, 7, 11, 14]. During the computation, the user-defined function is executed among vertices. In this model, the graph generally applies edge-cut partitions and distributes the vertices across different partitions.

- *Edge-centric model.* In this model, user-defined function is executed for each edge. In other words, edges are the first-class unit of computation and partitioning, and vertices that are attached to edges lying in different partitions are replicated and shared between those partitions. X-Stream [13] and Chaos [12] are two classical representatives of edge-centric based graph computing systems.
- *Subgraph-centric model.* Subgraph-centric model is more general than the above two models. When executing the graph algorithms, the user-defined function is oriented for a subgraph rather than a single vertex or edge. Tian et al. [19] first proposed "think like a graph" instead of "think like a vertex" abstraction after observing shortcomings in vertex-centric and edge-centric methods of graph processing. Later on, the works [16, 18] also use this model to speed up graph analysis tasks.
- *Others.* Besides the above three models, there are many other models. Block-centric model [23] partitions a large graph into blocks, and user-defined functions are then run over these blocks. Path-centric model [25] utilizes a set of tree-based partitions to model the graph and parallelly executes the user-defined functions over trees.

In this book, we concentrate on distributed graph computing systems with the vertex-centric computation model. In the next two subsections, we discuss two popular graph programming abstractions in the context of the vertex-centric model—vertex programming abstraction and gather–apply–scatter programming abstraction.

2.2 Vertex Programming Abstraction

The vertex programming abstraction treats a vertex as a primitive computation unit and allows users to concentrate on the logic of a vertex, which is called a vertex program, when implementing a distributed graph algorithm. This abstraction is first proposed in Pregel [11] and then is adopted by many other Pregel-like systems (e.g., Giraph [5], PAGE [15, 24], GPS [14], Apache Hama [17], GraphX [7], etc.).

Data Layout We first describe the data layout in Pregel-like systems. In order to distribute the graph among workers, Pregel-like systems apply edge-cut partitioning technique, which splits the vertex set of the graph into several non-intersection vertex subsets. Figure 2.4 is an example of partitioned four-vertices graph via edge-cut partitioning. In such partitioning strategy, each vertex is only assigned to a unique worker and it is the minimal computation logic unit. As a consequence, the computation on a graph is reduced to the computation on a vertex, which is abstracted as a vertex program. In a single iteration, every worker processes the stored subgraph vertex-by-vertex by calling the vertex program.

Vertex Program Figure 2.5 lists the classical APIs of a vertex program in Pregel-like systems. When user writes a graph algorithm, he/she just overrides the virtual

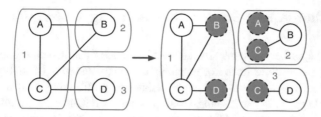

Fig. 2.4 Edge-cut partitioning. A four-vertices graph is partitioned into three subgraphs. The vertices with dashed outline mean the remote ones in each subgraph, respectively

```
1   //programing interfaces
2   template <typename VertexValue, typename EdgeValue, typename MessageValue>
3   class Vertex {
4     public:
5         virtual void Compute(MessageIterator* msgs) = 0;
6         const string& vertex_id() const;
7         int64 superstep() const;
8         const VertexValue& GetValue();
9         VertexValue* MutableValue();
10        OutEdgeIterator GetOutEdgeIterator();
11        void SendMessageTo(const string& dest_vertex, const MessageValue& message);
12        void VoteToHalt();
13  };
14
15  //implementation of Compute() method for PageRank
16  void Compute(MessageIterator* msgs){
17      newPR=0;
18      while(msgs->hasNext()){
19        newPR+=msgs->next()->getValue();
20      }
21      newPR=0.15/getOutEdgeIterator->size()+0.85*newPR;
22      SetValue(newPR);
23      OutEdgeIterator* dest=getOutEdgeIterator();
24      while(dest->hasNext()){
25        SendMessageTo(dest->next()->getValue(), newPR);
26      }
27  }
```

Fig. 2.5 APIs of vertex programming abstraction and a code sample of implementing PageRank algorithm, written in C++

method, *Compute*, which defines the main logic of the graph algorithm. The *SendMessageTo* method transmits the latest state of the vertex to a specified neighbor in the form of a message. The *VoteToHalt* method sets the vertex to be inactive when there is no more computation task. In the bottom of Fig. 2.5, we list a code sample of *Compute* method when implementing PageRank over Pregel-like systems. It first gets the latest PageRank values of the vertex's neighbors from the received messages (Lines 18–20) and then updates its own PageRank value (Lines 21–22). After the update, the latest PageRank value of the vertex is sent to the neighbor via *SendMessageTo* (Lines 24–25).

2.3   Gather–Apply–Scatter Programming Abstraction

Although Pregel-like systems advance performance of large-scale graph analysis, the system architecture does not consider the characteristics of real-world graph data. As mentioned before, they apply edge-cut partitioning to distribute graph among workers. This partitioning strategy tries to balance the number of graph vertices among each worker. However, in the real world, the degree of graph vertices satisfies the power-law distribution (i.e., $p(d) \propto d^{-\alpha}$), that is, most of the vertex degrees in the real world are very small, but a few of them are far greater than the average. On the basis of the balance of the number of vertices, the number of edges between the working nodes is highly skewed. However, for graph algorithms, the computation cost of a vertex is positively proportional to the vertex degree. Let us look at the PageRank implementation in Fig. 2.5, the number of sent messages (Lines 24–25) equals the out degree of the processed vertex. This makes a Pregel-like system cannot achieve an ideal load balance, and the performance of some workers will be much worse than average because of a few high-degree vertices.

Data Layout PowerGraph [6] applies a new fast approach to data layout for power-law graphs in distributed environments, thus effectively solving unbalanced workload problem. The new approach distributes edges of the graph among workers and then balances the load by balancing the number of edges between the workers. It causes a vertex that may have multiple copies in the system and select one copy to be central vertex and others are non-central vertices. We call such partitioning approach vertex-cut partitioning, which can parallelize the computation on a large vertex among multiple workers. Figure 2.6 illustrates a toy example where a four-vertices graph is partitioned into three subgraphs via vertex-cut. The vertices with dashed outline are the non-central vertices, respectively. For example, vertices C in partition 1 and 3 are non-central ones, while vertex C in partition 2 is the central one. In addition, the authors proved that any edge-cut strategy can be transformed into a better vertex-cut strategy efficiently.

Gather–Apply–Scatter (GAS) Abstraction The new data layout introduced by the vertex-cut strategy makes the data of a vertex be distributed among multiple

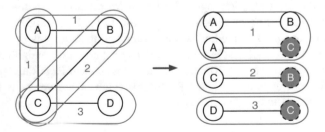

Fig. 2.6 Vertex-cut partitioning. A four-vertices graph partitioned via vertex-cut. The vertices with dashed outline mean non-central vertices, respectively

---

Algorithm 1 Vertex program in PowerGraph

Input: vertex u
1: **if** Cached state a_u is not empty **then**
2:     **foreach** v in u's neighbors **do**
3:         $a_u \leftarrow sum(a_u, gather(D_u, D_{(u,v)}, D_v))$
4:     **end foreach**
5: **end if**
6: $D_u \leftarrow apply(D_u, a_u)$
7: **foreach** v in u's neighbors **do**
8:     $(D_{(u,v)}, D_a) \leftarrow scatter(D_u, D_{(u,v)}, D_v)$
9:     **if** a_v and Δ_a are not empty **then**
10:        $a_v \leftarrow sum(a_v, \Delta_a)$
11:     **else**
12:        $a_v \leftarrow Empty$
13:     **end if**
14: **end foreach**

---

workers. The neighbors of a vertex are scattered on multiple workers, so the computation of a vertex can be parallelized among multiple workers. In order to ensure the correctness of the state of a single vertex after an iteration, PowerGraph gathers partial values from non-central vertices to the central vertex. Therefore, the vertex program defined in Pregel-like systems is no longer applicable. The PowerGraph proposes a new abstraction which factors individual vertex programs. The new abstraction is called gather–apply–scatter (GAS) model. Some other literatures [9] also name it edge-centric model, because an edge becomes the minimal computation unit in the context of vertex-cut partitioning. The APIs of GAS model are listed in Fig. 2.7.

Under the GAS abstraction, the execution flow of a vertex is divided into three stages: gather, apply, and scatter as well. Algorithm 1 illustrates the pseudo-procedure of the three stages:

1. In the gather phase, the vertices use the *Gather* and *Sum* methods to collect the state of neighbors. Different from Pregel-like systems, the *Gather* method collects neighbor states in parallel among multiple workers and then uses *Sum* method to cache the states in a_u and deliver them to the apply phase.
2. In the apply phase, the central vertex obtains the latest neighbor state by collecting all cached a_u and updates its own state by calling the *Apply* method and writes this update state back to all copies of the vertices, including central and non-central vertices.

Fig. 2.7 The APIs of GAS model, written in C++

```
1   interface GASVertexProgram(u) {
2       // Run on gather_nbrs(u)
3       gather(D_u, D_(u,v), D_v) → Accum
4       sum(Accum left, Accum right) → Accum
5       apply(D_u,Accum) → D_u^new
6       // Run on scatter_nbrs(u)
7       scatter(D_u^new ,D_(u,v),D_v) → (D_(u,v)^new, Accum)
8   }
```

3. In the scatter phase, multiple workers execute the *Scatter* method in parallel, passing the latest state to the adjacent edge of the vertex.

2.4   Workload-Aware Cost Model for Optimizations

To develop efficient graph algorithms on distributed graph computing systems, we need to investigate the factors that will influence the performance of the algorithms. Traditional optimization strategies are proposed from various aspects. In the data perspective, GPS [14] applies a technique called large adjacency-list partitioning (LALP), which distributively stores the adjacency lists of high-degree vertices across workers, thus parallelizing the neighborhood accessing. In the partition perspective, PowerGraph [6] uses vertex-cut partitioning to execute the computation of a node in parallel. In the architecture perspective, GPS introduces a master computation model to enable the master for central computation. Giraph [5] introduces multi-threading for CPU-bounded applications. Most of them do not **explicitly** consider the optimization problem in the view of workload. In this book, we design efficient distributed graph algorithms via explicitly modeling the workload and introducing a workload-aware cost model. The model clearly states different factors that affect the overall performance of the distributed graph computing systems. It is not only compatible with many other optimization techniques, but also can smartly guide us to find new optimization opportunities.

In this section, we first review the cost model for BSP-based systems, then introduce our workload-aware cost model over Pregel-like graph computing systems, and propose the principles of optimizations.

2.4.1   Cost Model Analysis

2.4.1.1   Cost Model of BSP-Based Systems

In Sect. 2.1, we introduce the execution flow of distributed graph computing systems. As shown in Fig. 2.2, in the context of BSP model, the total cost of a graph analysis task is the sum of the costs of supersteps and the synchronization costs between supersteps. The cost of a superstep is the highest cost among workers, and the cost of a worker in a superstep is determined by the computation cost and communication cost of the task. In the graph computing systems, computation and communication are executed concurrently, so the cost of a worker is the larger one between computation cost and communication cost. For example, in Fig. 2.2, the cost of the ith superstep is dominated by worker 2, and the cost of worker 2 is the computation cost.

Next we formally define the total cost of a graph analysis task. Assume that a graph analysis task is run on a graph computing system with P workers, and the

ith worker is denoted by W_i. The input graph $G = (V, E)$ is distributed stored on the P workers, and $V = \cup_{1 \leq i \leq P}\{V_{W_i}\}$, and $V_{W_i} \cap V_{W_j} = \emptyset, 1 \leq i, j \leq P$, where V_{W_i} is the graph vertex set stored in worker W_i. In addition, we assume that the computation cost and communication cost of a worker W_i are $C_p(W_i)$ and $C_m(W_i)$, respectively. Then the total cost of the ith worker is

$$C(W_i) = max\{C_p(W_i), C_m(W_i)\}. \qquad (2.1)$$

Based on the above notations, the total cost T of a graph analysis task can be expressed as

$$T = \sum_1^S max_{1 \leq i \leq P}\{C(W_i)\} + (S - 1)C_{sync}$$

$$\qquad (2.2)$$

$$= \sum_1^S max_{1 \leq i \leq P}\{max\{C_p(W_i), C_m(W_i)\}\} + (S - 1)C_{sync},$$

where S is the number of iterations, C_{sync} is the synchronization cost between iterations. Table 2.1 summarizes the notations frequently used in this book with regard to the workload.

Note that Eq. 2.2 clearly implies three main factors that heavily influence the total cost T of a graph analysis task. The factors are:

1. *The computation and communication cost of a worker, i.e., $C_p(W_i)$ and $C_m(W_i)$.* The cost of a worker is determined by the larger of $C_p(W_i)$ and $C_m(W_i)$. $C_p(W_i)$ and $C_m(W_i)$ reflect the workload of a single worker in the BSP model.
2. *The balance of workload distribution among workers.* The $max_{1 \leq i \leq P}$ function indicates that the workload of iterative graph analysis task is determined by the most expensive worker. In order to improve the parallelism of the algorithm, it is necessary to balance the workload among workers.

Table 2.1 Frequently used notations

Notation	Description
v	A vertex
$C_p(v)$	The computation cost of v
$C_m(v)$	The communication cost of v
W_i	The ith worker
$C(W_i)$	The total cost of worker W_i
$C_p(W_i)$	The total computation cost of worker W_i
$C_m(W_i)$	The total communication cost of worker W_i
T	The total cost of a graph analysis task
S	The number of supersteps (or iterations)
P	The number of workers
C_{sync}	The cost of synchronization between supersteps

3. *The number of iterations S*. In a distributed graph computing system (or BSP model), each iteration needs a time-consuming synchronization process. The larger the number of iterations S leads to the greater overhead of synchronization.

2.4.1.2   Cost Model in the Context of Graph

As mentioned before, $C_p(W_i)$ and $C_m(W_i)$ imply the workload of a single worker in the BSP model. Actually they are task-dependent and are determined by the complexity of the graph analysis task. Fortunately, in the context of graph and vertex-centric computation model, vertices in a graph are regarded as the minimal computing logic unit. In each iteration, different vertices are executed independently. Therefore, we can refine the representation of $C_p(W_i)$ and $C_m(W_i)$ in the form of vertices. In other words, the cost of a graph computation task is determined by the computation and communication of vertex programs. The computation cost $C_p(W_i)$ is the execution cost of the vertex program, while the communication cost $C_m(W_i)$ is determined by the amount of messages transmitted. Now, if the computation cost and communication cost of a vertex V are $C_p(v)$ and $C_m(v)$, respectively, then $C_p(W_i)$ and $C_m(W_i)$ can be expressed as

$$C_p(W_i) = \sum_{v \in V_{W_i}} C_p(v)$$

$$C_m(W_i) = \sum_{v \in V_{W_i}} C_m(v). \tag{2.3}$$

Thus, the total cost of a graph computing task can be rewritten as

$$T = \sum_{1}^{S} max_{1 \leq i \leq P}\{max\{ \sum_{v \in V_{W_i}} C_p(v), \sum_{v \in V_{W_i}} C_m(v)\}\} + (S - 1)C_{sync}. \tag{2.4}$$

Furthermore, the $C_p(v)$ and $C_m(v)$ are related to the complexity of a vertex program provided by the user and can be formally expressed as

$$C_p(v) = f(v, msg\_list),$$

$$C_m(v) = g(v, msg\_list), \tag{2.5}$$

where msg_list is the number of messages received by the vertex from the previous iteration. $f(v, msg\_list)$ is a function indicating the time cost of processing msg_list, while $g(v, msg\_list)$ is a function indicating the communication cost of transferring msg_list.

From Eq. 2.5, it can be seen that both the computation cost and the communication cost are related to the number of messages to be processed by the vertex and the complexity of processing the messages. This gives us a heuristic for graph algorithm optimization. In order to improve the performance of large-scale graph algorithm in graph computing systems, it is critical to design a simple vertex program and reduce the number of messages generated as much as possible.

Note that the above discussion assumes workers in a distributed system are homogeneous, i.e., each worker has the same computation and communication abilities. In the real environment, the workers are often heterogeneous [22]. We can revise the above cost model by carefully considering the capability (e.g., CPUs, memory, bandwidth, etc.) of computing nodes.

2.4.1.3 The Problem of Workload Balance

The distribution of workload among workers is an important factor that affects the overall cost of graph computing tasks. Suppose that if the total workload is distributed unevenly among the worker, for example, 80% of the workload of a task is handled by a single worker, while the remaining 20% workload is distributed among the other workers, then the parallelism of the task is extremely low and the efficiency cannot be improved significantly according to the Amdahl's law [2]. When distributing the workload among the workers, the more balance of the workload, the better the scalability of the algorithm. Actually, this is a classical load balance problem for any parallel algorithm. The problem directly affects the parallelism of a parallel algorithm. When the workload is heavily skewed, it is useless to increase the number of workers. More worse, the performance may be degraded because of the high communication cost among workers.

In order to quantitatively analyze the quality of workload distribution, we define the workload balance factor b_α as follows:

Definition (Workload Balance Factor b_α**)** *Given P workers, the workload of each worker* W_i *is* $C(W_i)$, *then* b_α *is*

$$b_\alpha = \frac{max_{1 \leq i \leq P}\{W_i\}}{min_{1 \leq i \leq P}\{W_i\}}. \tag{2.6}$$

According to the definition, $b_\alpha \geq 1$. If and only if $b_\alpha = 1$, the workload is perfectly balanced.

2.4.2 The Principles of Optimizations

With regard to the cost model analysis in Sect. 2.4.1, we can classify the factors that heavily influence the performance of a graph analytic task into two categories: **workload source** and **workload distribution**.

The workload source represents the factors that determine the total cost of a graph analytic task. The computation model, the complexity of graph algorithms, data characteristics are three major factors belonging to the workload source. Optimizing them will reduce the total cost of the graph analytic task.

The workload distribution represents the factors that affect the distribution of the workload among the workers, and it mainly contains data layout and physical resources. First, according to Eq. 2.4, the workload of a worker is related to the number of vertices it processed. Therefore, different data distribution brings different workload distribution. Second, it is mentioned in the previous analysis that the actual workload is related to physical resources equipped in a worker, so the workload distribution can also be fine-tuned by utilizing different amounts of physical resources.

To sum up, when developing an efficient graph algorithm over distributed graph computing systems, we follow the above workload-aware cost model, carefully consider the factors from workload source and workload distribution.

2.4.2.1 Optimizations with Respect to the Workload Source

Next we discuss the general ideas of optimizing the performance of graph analytic tasks with respect to the workload source and present them factor by factor.

First, from the computation model, we need to propose a new computation model with a more flexible programming interface. The new graph-oriented programming interface should not only ensure the simplicity of semantics, but also be flexible enough to allow professional algorithm designers or data analysts to deeply optimize their algorithms, so that they can achieve efficient complex graph algorithms.

Second, from the complexity of graph algorithms, the users/analysts need to fully study the essence of the algorithm and simplify (or approximate) the logic. For example, in order to reduce the number of iterations, we first analyze the variables and invariants in the algorithm, then by redesigning the algorithm, we can transform the remote update into the local update, and ultimately effectively reduce the number of iterations and synchronization costs. Meanwhile, such modification can also reduce the network traffic. In order to reduce the amount of computation and communication, an effective filtering mechanism can be established to decrease the number of messages generated.

Third, from the data characteristics, analysts need to carefully mine the data, understand the statistics of the data, and then design the optimization strategy combined with the data characteristics. This is because different data characteristics have different effects on the performance of the same algorithm. For example, the graph data in the real world generally meets the power-law distribution, and the cost estimation model should consider the skewed distribution to model the workload more accurate.

To conclude, the optimization strategies in terms of workload source can be summarized as follows:

1. Design a computation model for complex graph algorithm.
2. Study the computation logic of algorithms for optimization, e.g., building a filtering mechanism.
3. Consider the data characteristics for better cost estimation.

2.4.2.2 Optimizations with Respect to Workload Distribution

The main factors in workload distribution are data layout and physical resources. In distributed graph computing systems, there are two types of data, one is the input graph, and the other is the message data generated during runtime. In respect to the input graph, a good graph partitioning strategy by considering the characteristics of data can distribute the input graph evenly. However, when handling advanced graph algorithms, e.g., subgraph matching, the complexity of computation not only is affected by the graph data, but also is related to the message data. In respect to the message data, we should consider the data layout of messages online. This is because (1) the computation costs of different messages are different, we cannot find a static partition strategy based on graph data; (2) a single message of advanced graph algorithms may have multiple candidate workers to be processed, and the workloads on different workers are not the same, so we need to estimate the costs of messages dynamically and distribute them accordingly.

Physical resources also directly affect the workload distribution in the real computation environment. There are many researchers working on graph partitioning strategies for heterogeneous environment. However, these works still focus on the problem of data distribution. Besides that, we can also design the strategy of automatic resource allocation to ensure that the workers can complete the workload within certain time constraints when the workload distribution has been skewed, so as to achieve a good balance.

To sum up, in addition to the traditional load balancing strategy, we should further optimize them in the following two aspects:

1. Design workload-aware data distribution strategy to ensure the balance;
2. The automatic resource allocation strategy can ensure that the real computation costs of workers are almost the same.

References

1. V. Aggarwal, S. Sengupta, V. S. Sharma, and A. Santharam. A scalable master-worker architecture for PaaS clouds. In *2012 SC Companion: High Performance Computing, Networking Storage and Analysis*, pages 1268–1275, Nov 2012.

2. Gene M. Amdahl. Validity of the single processor approach to achieving large scale computing capabilities. In *Proceedings of the April 18–20, 1967, Spring Joint Computer Conference*, AFIPS '67 (Spring), pages 483–485, New York, NY, USA, 1967. ACM.
3. D. A. Bader and K. Madduri. Snap, small-world network analysis and partitioning: An open-source parallel graph framework for the exploration of large-scale networks. In *2008 IEEE International Symposium on Parallel and Distributed Processing*, pages 1–12, April 2008.
4. Albert Chan, Frank Dehne, and Ryan Taylor. CGMGRAPH/CGMLIB: Implementing and testing CGM graph algorithms on PC clusters and shared memory machines. In *Journal of HPCA*, pages 81–97, 2005.
5. Avery Ching, Sergey Edunov, Maja Kabiljo, Dionysios Logothetis, and Sambavi Muthukrishnan. One trillion edges: Graph processing at Facebook-scale. *Proc. VLDB Endow.*, 8(12):1804–1815, August 2015.
6. Joseph E. Gonzalez, Yucheng Low, Haijie Gu, Danny Bickson, and Carlos Guestrin. Powergraph: distributed graph-parallel computation on natural graphs. In *OSDI*, 2012.
7. Joseph E. Gonzalez, Reynold S. Xin, Ankur Dave, Daniel Crankshaw, Michael J. Franklin, and Ion Stoica. GraphX: Graph processing in a distributed dataflow framework. In *Proceedings of the 11th USENIX Conference on Operating Systems Design and Implementation*, OSDI'14, pages 599–613, Berkeley, CA, USA, 2014. USENIX Association.
8. Douglas Gregor and Andrew Lumsdaine. The parallel BGL: A generic library for distributed graph computations. In *POOSC*, 2005.
9. Safiollah Heidari, Yogesh Simmhan, Rodrigo N. Calheiros, and Rajkumar Buyya. Scalable graph processing frameworks: A taxonomy and open challenges. *ACM Comput. Surv.*, 51(3):60:1–60:53, June 2018.
10. Sungpack Hong, Hassan Chafi, Edic Sedlar, and Kunle Olukotun. Green-Marl: A DSL for easy and efficient graph analysis. In *Proceedings of the Seventeenth International Conference on Architectural Support for Programming Languages and Operating Systems*, ASPLOS XVII, pages 349–362, New York, NY, USA, 2012. ACM.
11. Grzegorz Malewicz, Matthew H. Austern, Aart J.C Bik, James C. Dehnert, Ilan Horn, Naty Leiser, and Grzegorz Czajkowski. Pregel: A system for large-scale graph processing. In *SIGMOD*, 2010.
12. Amitabha Roy, Laurent Bindschaedler, Jasmina Malicevic, and Willy Zwaenepoel. Chaos: Scale-out graph processing from secondary storage. In *Proceedings of the 25th Symposium on Operating Systems Principles*, SOSP '15, pages 410–424, New York, NY, USA, 2015. ACM.
13. Amitabha Roy, Ivo Mihailovic, and Willy Zwaenepoel. X-stream: Edge-centric graph processing using streaming partitions. In *Proceedings of the Twenty-Fourth ACM Symposium on Operating Systems Principles*, SOSP '13, pages 472–488, New York, NY, USA, 2013. ACM.
14. Semih Salihoglu and Jennifer Widom. GPS: A graph processing system. In *SSDBM*, 2013.
15. Y. Shao, B. Cui, and L. Ma. PAGE: A partition aware engine for parallel graph computation. *IEEE Transactions on Knowledge and Data Engineering*, 27(2):518–530, Feb 2015.
16. Yingxia Shao, Lei Chen, and Bin Cui. Efficient cohesive subgraphs detection in parallel. In *Proc. of ACM SIGMOD Conference*, pages 613–624, 2014.
17. Kamran Siddique, Zahid Akhtar, Yangwoo Kim, Young-Sik Jeong, and Edward J. Yoon. Investigating Apache Hama: A bulk synchronous parallel computing framework. *J. Supercomput.*, 73(9):4190–4205, September 2017.
18. Yogesh Simmhan, Alok Kumbhare, Charith Wickramaarachchi, Soonil Nagarkar, Santosh Ravi, Cauligi Raghavendra, and Viktor Prasanna. Goffish: A sub-graph centric framework for large-scale graph analytics. In Fernando Silva, Inês Dutra, and Vítor Santos Costa, editors, *Euro-Par 2014 Parallel Processing*, pages 451–462, Cham, 2014. Springer International Publishing.
19. Yuanyuan Tian, Andrey Balmin, Severin Andreas Corsten, Shirish Tatikonda, and John McPherson. From "think like a vertex" to "think like a graph". *Proc. VLDB Endow.*, 7(3):193–204, November 2013.
20. Leslie G. Valiant. A bridging model for parallel computation. *Commun. ACM*, 33(8):103–111, August 1990.

21. David W. Walker, David W. Walker, Jack J. Dongarra, and Jack J. Dongarra. Mpi: A standard message passing interface. *Supercomputer*, 12:56–68, 1996.
22. N. Xu, B. Cui, L. Chen, Z. Huang, and Y. Shao. Heterogeneous environment aware streaming graph partitioning. *IEEE Transactions on Knowledge and Data Engineering*, 27(6):1560–1572, June 2015.
23. Da Yan, James Cheng, Yi Lu, and Wilfred Ng. Blogel: A block-centric framework for distributed computation on real-world graphs. *PVLDB*, 7(14):1981–1992, 2014.
24. Shao Yingxia, Yao Junjie, Cui Bin, and Ma Lin. PAGE: A partition aware graph computation engine. In *CIKM*, pages 823–828, 2013.
25. P. Yuan, W. Zhang, C. Xie, H. Jin, L. Liu, and K. Lee. Fast iterative graph computation: A path centric approach. In *SC '14: Proceedings of the International Conference for High Performance Computing, Networking, Storage and Analysis*, pages 401–412, Nov 2014.

Chapter 3
Partition-Aware Graph Computing System

Abstract Graph partition quality affects the overall performance of distributed graph computing systems. The quality of a graph partition is measured by the balance factor and edge cut ratio. A balanced graph partition with small edge cut ratio is generally preferred since it reduces the high network communication cost. However, through an empirical study on Giraph, we find that the performance over well partitioned graph might be even two times worse than simple random partitions. The reason is that the systems only optimize for the simple partition strategies and cannot efficiently handle the increasing workload of local message processing when a high-quality graph partition is used. In this chapter, we introduce a novel partition-aware graph computing system named PAGE, which equips a new message processor and a dynamic concurrency control model. The new message processor concurrently processes local and remote messages in a unified way. The dynamic model adaptively adjusts the concurrency of the processor based on the online statistics. The experimental studies demonstrate the superiority of PAGE over the graph partitions with various qualities.

3.1 Introduction

With the development of Web 2.0 and mobile internet, there emerge a wide array of applications which analyze massive graphs, like community detection [14], linkage analysis [6, 12], graph pattern matching [16], and machine learning models [2]. In front of the large graphs, distributed processing becomes the de-facto technique for graph computing framework. Many distributed graph computing systems are proposed like Pregel [17], Giraph [7], GPS [21], and GraphLab [15]. In these systems, a graph algorithm consists of several supersteps which are separated by synchronization barriers. In each superstep, every active vertex simultaneously updates its status based on the neighbors' messages from the previous superstep, and then sends the latest status as a message to the neighbors.

Considering the limited number of workers in practice, a single worker in the systems usually stores a subgraph at local, and computes the local active vertices one by one. The computations among the workers are in parallel. Therefore, graph

© Springer Nature Singapore Pte Ltd. 2020
Y. Shao et al., *Large-scale Graph Analysis: System, Algorithm and Optimization*,
Big Data Management, https://doi.org/10.1007/978-981-15-3928-2_3

partition is one of critical components that influence the performance of graph computing systems. It aims to partition the original graph into several subgraphs, which are of about the same size and only contain few cross edges among them. A high-quality graph partition means there are few edges between different subgraphs while each subgraph has similar size, and we also call it a good balanced graph partition. We define **edge cut** as the ratio of the edges crossing different subgraphs to the total edges. A good balanced partition has a small edge cut, which reduces the expensive communication cost among workers, thus improving the efficiency of systems. Furthermore, the balance property ensures each subgraph incurs similar computation workload.

However, in real scenarios, a good balanced graph partition even leads to a reduction of the overall performance in existing systems. In Fig. 3.1, we present the performance of PageRank over six different partition schemes of a large web graph, and obviously the overall cost of PageRank per iteration increases when the qualities of different graph partitions improve. For example, when the edge cut is about 3.48% in METIS, the performance is about two times worse than the one in simple random partition scheme where edge cut is 98.52%. The results imply that the distributed graph computing systems may not benefit from the high-quality graph partition.

Figure 3.1b also illustrates the local communication cost and sync remote communication cost (defined in Sect. 3.2). We can see that, when the edge cut ratio decreases, the sync remote communication cost is reduced as expected. However, the local communication cost increases fast unexpectedly, which causes the downgrade of overall performance. This abnormal phenomenon implies the local message processing becomes a bottleneck in the system and dominates the overall cost when the workload of local message processing increases.

Actually, most of the existing distributed graph computing systems are unaware of such influence caused by the underlying graph partitions, and ignore the increasing workload of local message processing when the quality of partition scheme is

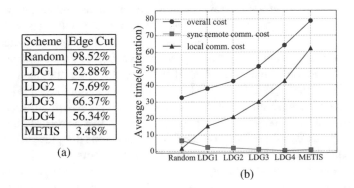

Scheme	Edge Cut
Random	98.52%
LDG1	82.88%
LDG2	75.69%
LDG3	66.37%
LDG4	56.34%
METIS	3.48%

(a)

(b)

Fig. 3.1 PageRank on various partitions over web graph uk-2007-05-u. The detailed experiment setup is described in Sect. 3.4. (**a**) Partition quality. (**b**) Time cost per iteration

improved. These systems handle the local messages and remote messages different. They only optimize the processing of remote messages, since the cost of remote message processing dominates the total cost with random graph partition scheme. In other words, a balanced graph partition is not corresponding to the balanced workload. There is a simple extension of centralized message buffer to process local and remote incoming messages all together [21], however, this extension still cannot help the distributed graph computing systems effectively utilize the benefit of high-quality graph partitions.

To solve the above challenges, we introduce a novel graph computing engine, called partition-aware graph computing engine (PAGE). PAGE equips with two new components to efficiently leverage the different qualities of the underlying partition schemes. First, in a PAGE's worker, communication module is extended with a new dual concurrent message processor. The message processor concurrently handles both local and remote incoming messages in a unified way, thus accelerating the message processing. Furthermore, the concurrency of the message processor is automatically tunable according to the online statistics of the system. Second, a partition-aware module is added in each worker to monitor the partition-related statistics and adjust the concurrency of the message processor adaptively to fit the online dynamic workload.

To estimate the concurrency for the dual concurrent message processor rationally, we introduce the dynamic concurrency control model (DCCM). Since the message processing pipeline satisfied the producer–consumer model, heuristic rules are proposed on the basis of the producer–consumer constraints. With the help of DCCM, PAGE is able to allocate sufficient message process units to handle current workload and each message process unit has similar workload. Finally, PAGE can adaptively utilize the different qualities of the underlying graph partitions. We build a prototype of PAGE on top of Giraph (version 0.2.0). The experimental results demonstrate that the PAGE can improve the efficiency of both stationary and non-stationary graph algorithms on graph partitions with various qualities.

The remains of this chapter are organized as follows. In Sect. 3.2, we describe the characteristics of message processing in graph computation systems. In Sect. 3.3, we elaborate the framework of PAGE. The experimental results are described in Sect. 3.4. Finally, we conclude this chapter in Sect. 3.5.

3.2 Message Processing in Pregel-Like Systems

In Pregel-like graph computing systems, vertices exchange their status through message passing. When the vertex sends a message, the worker first determines whether the destination vertex of the message is owned by a remote worker or the local worker. In the remote case, the message is buffered first. When the buffer size exceeds a certain threshold, the largest one is asynchronously flushed, delivering each to the destination as a single message. In the local case, the message is directly placed in the destination vertex's incoming message queue [17].

Therefore, the communication cost in a single worker is split into local communication cost and remote communication cost. Combining the computation cost, the overall cost of a worker has three components. Computation cost, denoted by t_{comp}, is the cost of execution of vertex programs. Local communication cost, denoted by t_{comml}, represents the cost of processing messages generated by the worker itself. Remote communication cost, denoted by t_{commr}, includes the cost of sending messages to other workers and waiting for them processed. We use the cost of processing remote incoming messages at local to approximate the remote communication cost. There are two reasons for adopting such an approximation. First, the difference between two costs is the network transferring cost, which is relatively small compared to remote message processing cost. Second, the waiting cost of the remote communication cost is dominated by the remote message processing cost.

The workload of local (remote) message processing determines the local (remote) communication cost. The graph partition influences the workload distribution of local and remote message processing. A high-quality graph partition, which is balanced and has small edge cut ratio, usually leads to that the local message processing workload is higher than the remote message processing workload.

3.2.1 The Influence of Graph Algorithms

In reality, besides the graph partition, the actual workload of message processing in an execution instance is related to the characteristics of graph algorithms as well.

Here we follow the graph algorithm category introduced in [11]. On basis of the communication characteristics of graph algorithms when running on a vertex-centric graph computation system, they are classified into stationary graph algorithms and non-stationary graph algorithms. The **stationary graph algorithms** have the feature that all vertices send and receive the same distribution of messages between supersteps, like PageRank, diameter estimation [9]. In contrast, the destination or size of the outgoing messages changes across supersteps in the **non-stationary graph algorithms**. For example, traversal-based graph algorithms, e.g., breadth first search and single source shortest path are the non-stationary ones.

In the stationary graph algorithms, every vertex has the same behavior during the execution, so the workload of message processing only depends on the underlying graph partition. When a high-quality graph partition is applied, the local message processing workload is higher than the remote one, and vice versa. The high-quality graph partition helps improve the overall performance of stationary graph algorithms, since processing local messages is cheaper than processing remote messages, which has a network transferring cost.

For the traversal-based graph algorithms belonging to the non-stationary category, it is also true that the local message processing workload is higher than the remote one when a high-quality graph partition is applied. Because the high-quality graph partition always clusters a dense subgraph together, which is traversed in successive supersteps. However, the high-quality graph partition cannot guarantee a better overall performance for the non-stationary ones, because of the workload imbalance of the algorithm itself. This problem can be solved by techniques in [11, 24].

We focus on the efficiency of a worker when different quality graph partitions are applied. The systems finally achieve better performance by improving the performance of each worker and leave the workload imbalance to the dynamic repartition solutions. The next subsection will reveal the drawback in the existing systems when handling different quality graph partitions.

3.2.2 The Total Cost of a Worker

As mentioned before, the cost of a worker has three components. Under different designs of the communication module, there are several combinations of above three components to determine the overall cost of a worker. Figure 3.2 lists two possible combinations and illustrates fine-grained cost ingredients as well. Components in a single bar mean that their costs are additive because of the sequential processing. The overall cost equals the highest one among these independent bars.

The cost combination in Giraph is illustrated in Fig. 3.2a. The computation cost and local communication cost are in the same bar, as Giraph directly processes local message processing during the computation. The sync remote communication cost, $t_{syncr} = t_{commr} - t_{comp} - t_{comml}$, is the cost of waiting for the remote

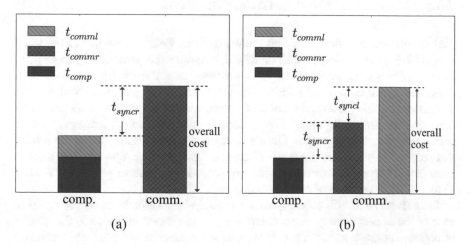

Fig. 3.2 Different combinations of computation cost, local/remote communication cost. The arrows indicate the ingredients of overall cost. (**a**) Combination in Giraph. (**b**) Combination in PAGE

incoming message processing to be accomplished after the computation and local message processing finished. This type of the combination processes local incoming messages and remote incoming messages unequally, and the computation might be blocked by processing local incoming messages. When the workload of processing local incoming messages increases, the performance of a worker degrades severely. This is the main cause that Giraph suffers from a good balanced graph partition which is presented in Sect. 3.1.

3.3 PAGE: A Partition-Aware System for Large-Scale Graph Analysis

PAGE, which stands for **P**artition **A**ware **G**raph computation **E**ngine, is designed to support different graph partition qualities and maintain high performance by an adaptively tuning mechanism and new cooperation methods. Figure 3.3a illustrates the architecture of PAGE. Similar to the majority of existing parallel graph computation systems, PAGE follows the master–worker paradigm. The computing graph is partitioned and distributively stored among workers' memory. The master is responsible for aggregating global statistics and coordinating global synchronization. The novel worker in Fig. 3.3b is equipped with an enhanced communication module and a newly introduced partition-aware module. Thus the workers in PAGE can be aware of the underlying graph partition information and optimize the graph computation task.

3.3.1 Graph Algorithm Execution in PAGE

The main execution flow of graph computation in PAGE is similar to the other Pregel-like systems. However, since PAGE integrates the partition-aware module, there exist some extra works in each superstep and the modified procedure is illustrated in Algorithm 1. At the beginning, the DCCM in partition-aware module calculates suitable parameters based on metrics from the previous superstep, and then updates the configurations (e.g., concurrency, assignment strategy) of dual concurrent message processor. During a superstep, the monitor tracks the related statistics of key metrics in the background. The monitor updates key metrics according to these collected statistics and feeds up to date values of the metrics to the DCCM at the end of each superstep.

Note that PAGE will reconfigure the message processor in every superstep in case of the non-stationary graph algorithms. As discussed in Sect. 3.2, the quality of underlying graph partition may change between supersteps for the non-stationary graph algorithms with dynamic workload balance strategy. Even if the static graph partition strategy is used, the variable workload characteristics of non-stationary

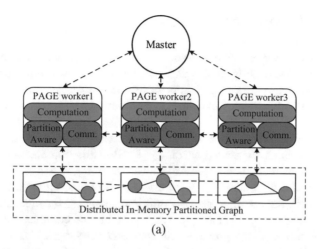

PAGE Worker

Fig. 3.3 The framework of PAGE. (**a**) Architecture. (**b**) Partition aware worker

graph algorithms require the reconfigurations across supersteps. In summary, PAGE can adapt to different workloads caused by the underlying graph partition and the graph algorithm itself.

Algorithm 1 Procedure of a superstep in PAGE

1: *DCCM reconfigures dual concurrent message processor parameter.*
2: **foreach** active vertex v in partition P **do**
3:     call vertex program of v;
4:     send messages to the neighborhood of v;
5:     /* *monitor tracks related statistics in the background.* */
6: **end foreach**
7: synchronization barrier
8: *monitor updates key metrics, and feeds to the DCCM*

3.3.2   Dual Concurrent Message Processor

The dual concurrent message processor is the core of the enhanced communication model, and it concurrently processes local and remote incoming messages in a unified way. With proper configurations for this new message processor, PAGE can efficiently deal with incoming messages over various graph partitions with different qualities.

As discussed in Sect. 3.2, messages are delivered in block, because the network communication is an expensive operation [8]. But this optimization raises extra overhead in terms that when a worker receives incoming message blocks, it needs to parse them and dispatches extracted messages to the specific vertex's message queue. In PAGE, the **message process unit** is responsible for this extra overhead, and it is a minimal independent process unit in the communication module. A remote (local) message process unit only processes remote (local) incoming message blocks. The **message processor** is a collection of message process units. The remote (local) message processor only consists of remote (local) message process units. Figure 3.4 illustrates the pipeline that the message process unit handles the overhead.

According to the cost analysis in Sect. 3.2.2, we can see that a good solution is to decouple the local communication cost from the computation cost, as the computation will not be blocked by any communication operation. Besides, the communication module can take over both local and remote communications, which makes it possible to process local and remote messages in a unified way. Furthermore, the incoming message blocks are concurrently received from both local and remote sources. It is better to process the local and remote incoming messages separately. These two observations help us design a novel message processor, which consists of a local message processor and a remote message processor. This novel message processor design leads to the cost combination in Fig. 3.2b. The sync local communication cost, $t_{syncl} = t_{comml} - t_{commr}$, is similar to the sync remote communication cost. It is the cost of waiting for the local incoming message processing to be accomplished after all the remote messages have been processed.

Moreover, in parallel graph computation systems, the incoming messages are finally appended to the vertex's message queue, so different vertices can be

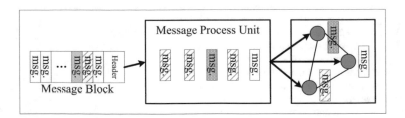

Fig. 3.4 Message processing pipeline in PAGE

easily updated concurrently. Taking this factor into consideration, we deploy the concurrent message process units at the internal of local and remote message processor. Therefore, both local and remote message processors can concurrently process incoming message blocks, and local and remote incoming messages are processed in a unified way.

In summary, the new message processor consists of a local and a remote message processor, respectively. This is the first type of the concurrency in the message processor. The second type of the concurrency is at the internal of local and remote message processor. This explains the reason we name this new message processor as dual concurrent message processor.

3.3.3 Partition-Aware Concurrency Control Model

The concurrency of dual concurrent message processor heavily affects the performance of PAGE. But it is expensive and also challenging to determine a reasonable concurrency ahead of real execution without any assumption [19]. Therefore, PAGE needs a mechanism to adaptively tune the concurrency of the dual concurrent message processor, thus being consistent with the dynamic workloads and the underlying graph partition scheme. The mechanism is named dynamic concurrency control model, DCCM for short.

In PAGE, the concurrency control problem can be modeled as a typical producer–consumer scheduling problem, where the computation phase generates messages as a producer, and message process units in the dual concurrent message processor are the consumers. Therefore, the producer–consumer constraints [22] should be satisfied when solving the concurrency control problem.

For the PAGE situation, the concurrency control problem arises consumer constraints. Since the behavior of producers is determined by the graph algorithms, PAGE only requires to adjust the consumers to satisfy the constraints (behavior of graph algorithms), which are stated as follows.

First, PAGE provides sufficient message process units to make sure that new incoming message blocks can be processed immediately and do not block the whole system. Meanwhile, no message process unit is idle.

Second, the assignment strategy of these message process units ensures that each local/remote message process unit has balanced workload since the disparity can seriously destroy the overall performance of parallel processing.

Above requirements derive two heuristic rules:

Rule 1: **Ability lower bound:** the message processing ability of all the message process units should be no less than the total workload of message processing.

Rule 2: **Workload balance ratio:** the assignment of total message process units should satisfy the workload ratio between local and remote message processing.

Table 3.1 Frequently used notations

Symbols	Description
er	Edge cut ratio of a local partition
p	Quality of network transfer
s_p	Message processing speed
s_g	Message generating speed
s_{rg}	Speed of remote message generation
s_l	Incoming speed of local messages
s_r	Incoming speed of remote messages
n_{mp}	Number of message process units
n_{rmp}	Number of remote message process units
n_{lmp}	Number of local message process units
t_{comp}	Computation cost
t_{comm}	Communication cost
t_{sycnr}	Cost of syncing remote communication
t_{sycnl}	Cost of syncing local communication

Following subsections first introduce the mathematical formulation of DCCM, and then discuss how DCCM mitigates the influences of various graph partition qualities. At last, we present the implementation of DCCM. Table 3.1 summarizes the frequently used notations in the following analysis.

3.3.3.1 Mathematical Formulation of DCCM

In PAGE, DCCM uses a set of general heuristic rules to determine the concurrency of dual concurrent message processor. The workload of message processing is the number of incoming messages, and it can be estimated by the incoming speed of messages. Here we use s_l and s_r to denote the incoming speed of local messages and the incoming speed of remote messages, respectively. The ability of a single message processing unit is the speed of processing incoming messages, s_p.

On basis of aforementioned two heuristic rules, the following equations must hold.

$$n_{mp} \times s_p \geq s_l + s_r, \quad \text{Rule 1}$$
$$\frac{n_{lmp}}{n_{rmp}} = \frac{s_l}{s_r}. \quad \text{Rule 2} \tag{3.1}$$

where n_{mp} stands for the total number of message process units, n_{lmp} represents the number of local message process units, and n_{rmp} is the number of remote message process unit. Meanwhile, $n_{mp} = n_{lmp} + n_{rmp}$.

Solving Eq. 3.1 yields

$$n_{lmp} \geq \frac{s_l}{s_p},$$

$$n_{rmp} \geq \frac{s_r}{s_p}.$$

(3.2)

Finally, DCCM can estimate the concurrency of local message processor and the concurrency of remote message processor separately, if the metrics s_l, s_r, s_p are known. The optimal concurrency reaches when DCCM sets $n_{lmp} = \lceil \frac{s_l}{s_p} \rceil$ and $n_{rmp} = \lceil \frac{s_r}{s_p} \rceil$. Because, at this point, PAGE can provide sufficient message process units, and it consumes minimal resources as well.

3.3.3.2 Adaptiveness on Various Graph Partitions

Section 3.2 has analyzed that the workload of message processing is related to both graph partition and graph algorithms. In this section, we explain the reason that previous DCCM can adaptively tune the concurrency of dual concurrent message processor when the underlying graph partition quality is changed.

Before the detailed discussion, we first give three metrics.

1. **Edge cut ratio of a local partition.** It is the ratio between cross edges and total edges in a local graph partition. It is denoted as er. This metric judges the quality of graph partitioning in a worker. The higher ratio indicates the lower quality.
2. **Message generating speed.** It represents the overall generating velocity of all outgoing messages in a worker. This metric implies the total workload for a worker. We denote it as s_g.
3. **Quality of network transfer.** This reflects the degree of network influence to the message generation speed s_g. When the generated messages are sent to a remote worker, the speed of generated messages is cut-off in the view of the remote worker. This is caused by the factor that network I/O operation is slower than local operation. The quality is denoted as $p \in (0, 1)$. The larger p indicates the better network environment. In addition, we can define the equivalent speed of remote messages' generation as $s_{rg} = s_g \times p$.

Now we proceed to reveal the relationship between DCCM and the underlying graph partition. Following analysis is based on the assumption that the stationary graph algorithms are ran on a certain partition. Because stationary graph algorithms have predictable communication feature.

The local incoming messages are the one whose source vertex and destination vertex belong to the same partition. Thus, the incoming speed of local message, s_l, is the same as $s_g \times (1 - er)$, which stands for the generating speed of local messages.

Similarly, s_r equals $s_{rg} \times er$. Then Eq. 3.1 can be rewrote as

$$n_{mp} \times s_p = s_g \times (1 - er) + s_{rg} \times er,$$

$$\frac{n_{lmp}}{n_{rmp}} = \frac{s_g \times (1 - er)}{s_{rg} \times er} = \frac{1 - er}{p \times er}. \tag{3.3}$$

Solving n_{mp}, n_{lmp}, n_{rmp} from Eqs. 3.3, we can have the following results

$$n_{mp} = \frac{s_g}{s_p}(1 - (1 - p) \times er), \tag{3.4}$$

$$n_{lmp} = n_{mp} \times \frac{1 - er}{p \times er + (1 - er)}, \tag{3.5}$$

$$n_{rmp} = n_{mp} \times \frac{p \times er}{p \times er + (1 - er)}. \tag{3.6}$$

From Eqs. 3.4, 3.5, and 3.6, we have following observations that indicate correlated relationships between graph partition and the behavior of DCCM when running stationary graph algorithms on PAGE.

First, PAGE needs more message process units with the quality growth of graph partitioning, but the upper bound still exists. This is derived from the fact that, in Eq. 3.4, the n_{mp} increases while er decreases, since the p is fixed in a certain environment. However, the conditions, $0 < p < 1$ and $0 < er < 1$, always hold, so that n_{mp} will not exceed s_g/s_p. Actually, not only the parameters s_g, s_p dominate the upper bound of total message process units, but also p heavily affects the accurate total number of message process units under various partitioning quality during execution.

The accurate total number of message process units is mainly affected by s_g and s_p, while er only matters when the network is really in low quality. Usually in a high-end network environment where p is large, the term $(1 - p) \times er$ is negligible in spite of er, and then the status of whole system (s_g, s_p) determines the total number of message process units. Only in some specific low-end network environments, the graph partitioning quality will severely affect the decision of total number of message process units.

Unlike the total number of message process units, the assignment strategy is really sensitive to the parameter er. From Eq. 3.3, we can see that the assignment strategy is heavily affected by $(1-er)/er$, as the value of p is generally fixed for a certain network. Lots of existing systems, like Giraph, do not pay enough attention to this phenomenon and suffer from high quality graph partitions. Our DCCM can easily avoid the problem by handling online assignment based on Eqs. 3.5 and 3.6.

Finally, when the non-stationary graph algorithms are ran on PAGE, the graph partition influence to the DCCM is similar as before. The difference is that the edge cut ratio of a local partition is only a hint, not the exact ratio for local and remote incoming message distribution. Because the unpredictable communication features of non-stationary graph algorithms cannot guarantee that a lower er leads to higher workload of local message processing. However it does for many applications in reality, such as traversal-based graph algorithms.

3.3.3.3 Implementation of DCCM

Given the heuristic rules and characteristic discussion of the DCCM, we proceed to present its implementation issues within the PAGE framework. To incorporate the DCCM's estimation model, PAGE is required to equip a monitor to collect necessary information in an online way. Generally, the monitor needs to maintain three high-level key metrics: s_p, s_l, s_r. However, there is a problem that it is difficult to measure accurate s_p. Because the incoming message blocks are not continuous and the granularity of time in the operating system is not precise enough, it is hard for the DCCM to obtain an accurate time cost of processing these message blocks. In the end, it leads to an inaccurate s_p.

Therefore, we introduce a DCCM in incremental fashion based on the original DCCM. Recall the cost analysis in Sect. 3.2, we can materialize the workload through multiplying the speed and the corresponding cost. Equation 3.2 can be represented as follows (use "$=$" instead of "$\geq$").

$$s_l \times t_{comp} = s_p \times n'_{lmp} \times (t_{comp} + t_{syncr} + t_{syncl}) \tag{3.7}$$

$$s_r \times t_{comp} = s_p \times n'_{rmp} \times (t_{comp} + t_{syncr}) \tag{3.8}$$

where n'_{lmp} and n'_{rmp} are the concurrency of local and remote message processor in current superstep, respectively.

The estimated concurrency of the local and remote message processor for the next superstep can be

$$n_{lmp} = \frac{s_l}{s_p} = n'_{lmp} \times (1 + \frac{t_{syncr} + t_{syncl}}{t_{comp}}) \tag{3.9}$$

$$n_{rmp} = \frac{s_r}{s_p} = n'_{rmp} \times (1 + \frac{t_{syncr}}{t_{comp}}) \tag{3.10}$$

Finally, the new DCCM can estimate the latest n_{lmp} and n_{rmp} based on the previous values and the corresponding time cost $t_{comp}, t_{syncl}, t_{syncr}$. The monitor is only responsible to track three cost-related metrics. As the monitor just records the time cost without any additional data structures or complicated statistics, it brings the negligible overhead. In our PAGE prototype system implementation, we apply the DCCM with incremental fashion to automatically determine the concurrency of the system.

3.4 Experiments

We have implemented the PAGE prototype on top of an open-source Pregel-like graph computation system, Giraph [7]. To test its performance, we conducted extensive experiments and demonstrated the superiority of our proposal. The following section describes the experimental environment, datasets, baselines, and evaluation metrics. The detailed experiments evaluate the effectiveness of DCCM, the advantages of PAGE compared with other methods, and the performance of PAGE on various graph algorithms.

3.4.1 Experimental Setup

All experiments are ran on a cluster of 24 nodes, where each physical node has an AMD Opteron 4180 2.6 Ghz CPU, 48 GB memory, and a 10 TB disk RAID. Nodes are connected by 1 Gbt routers. Two graph datasets are used: uk-2007-05-u [4, 5] and livejournal-u [3, 13]. The uk-2007-05-u is a web graph, while livejournal-u is a social graph. Both graphs are undirected ones created from the original release by adding reciprocal edges and eliminating loops and isolated nodes. Table 3.2 summarizes the meta-data of these datasets with both directed and undirected versions.

Graph Partition Strategies We partition the large graphs with three strategies: Random, METIS, and linear deterministic greedy (LDG). METIS [10] is a popular off-line graph partition packages, and LDG [23] is a well-known stream-based graph partition solution. The uk-2007-05-u graph is partitioned into 60 subgraphs, and livejournal-u graph is partitioned into 2, 4, 8, 16, 32, 64 partitions, respectively. Balance factors of all these partitions do not exceed 1%, and edge cut ratios are list in Fig. 3.1a and Table 3.3. The parameter setting of METIS is the same as METIS-balanced approach in GPS [21].

Furthermore, in order to generate various partition qualities of a graph, we extend the original LDG algorithm to an iterative version. The iterative LDG partitions the graph based on previous partition result, and gradually improves the partition quality in every following iteration. We name the partition result from iterative LDG as LDGid, where a larger id indicates the higher quality of graph partitioning and the more iterations executed.

Table 3.2 Graph dataset information

Graph	Vertices	Edges	Directed
uk-2007-05	105, 896,555	3,738,733,648	Yes
uk-2007-05-u	105,153,952	6,603,753,128	No
livejournal	4,847,571	68,993,773	Yes
livejournal-u	4,846,609	85,702,474	No

Table 3.3 Partition quality of livejournal-u

# partition	LDG (%)	Random (%)	METIS (%)
2 partitions	20.50	50.34	6.46
4 partitions	34.24	75.40	15.65
8 partitions	47.54	87.86	23.54
16 partitions	52.34	94.04	28.83
32 partitions	55.55	97.08	32.93
64 partitions	57.36	98.56	36.14

Baselines Throughout all experiments, we use two baselines for the comparison with PAGE.

The first one is Giraph. However, as we notice from Fig. 3.2a that the local message processing and computation run serially in the default Giraph. This cost combination model is inconsistent with our evaluation. We modify it to asynchronously process the local messages, so that Giraph can concurrently run computation, local message processing, and remote message processing. In the following experiments, **Giraph** refers to this modified Giraph version. Note that, this modification will not decrease the performance of Giraph.

The other one is derived from the technique used in GPS. One optimization in GPS, applies a centralized message buffer and sequentially processes incoming messages without synchronizing operations, which decouples the local message processing from the computation and treats the local and remote message equivalently. We implement this optimization on the original Giraph and denote it as **Giraph-GPSop**.

Metrics for Evaluation We use the following metrics to evaluate the performance of a graph algorithm on a graph computation system.

- **Overall cost.** It indicates the whole execution time of a graph algorithm when running on a computation system. Due to the property of concurrent computation and communication model, this metric is generally determined by the slower one between the computation and communication.
- **Sync remote communication cost.** It presents the cost of waiting for all I/O operations to be accomplished successfully after the computation finished. This metric reveals the performance of remote message processing.
- **Sync local communication cost.** It means the cost of waiting for all local messages to be processed successfully after syncing remote communication. This metric indicates the performance of local message processing.

All three metrics are measured by the average time cost per iteration. The relationship among these metrics can be referred to Fig. 3.2.

3.4.2　Evaluation on the Concurrency Control Model

Dynamic concurrency control model (DCCM) is the key component of PAGE to determine the concurrency for dual concurrent message processor, balance the workload for both remote and local message processing as well, and hence improve the overall performance. In this section, we demonstrate the effectiveness of DCCM through first presenting the concurrency automatically chosen by DCCM based on its estimation model, and then showing that these chosen values are close to the manually tuned good parameters. Finally, we also show DCCM converges efficiently to estimate a good concurrency for dual concurrent message processor.

3.4.2.1　Concurrency Determined by DCCM

Figure 3.5 shows the concurrency of dual concurrent message processor automatically determined by the DCCM, when running PageRank on uk-2007-05-u graph. The variation of concurrency on various graph partition schemes is consistent with the analysis in Sect. 3.3.3.2.

In Fig. 3.5, there are two meaningful observations. First, the total number of message process units, n_{mp}, is always equal to seven across six different partition schemes. This is because the cluster has a high-speed network and the quality of network transfer, p, is high as well. Second, with the decrease of edge cut ratio (from left to right in Fig. 3.5), the n_{lmp} decided by the DCCM increases smoothly to handle the growing workload of local messages processing, and the selected n_{rmp} goes oppositely. According to above parameters' variations across different graph partition schemes, we also conclude that the assignment strategy is more sensitive to the edge cut ratio than the total message process units n_{mp}.

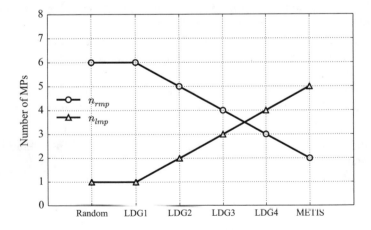

Fig. 3.5　Assignment strategy on different partition schemes in PAGE

3.4.2.2 Results by Manual Tuning

Here we conduct a series of manual parameter tuning experiments to discover the best configurations for the dual concurrent message processor when running PageRank on uk-2007-05-u in practice. Hence, it helps verify that the parameters determined by DCCM are effective.

We tune the parameters n_{lmp} and n_{rmp} one by one on a specific partition scheme. When one variable is tuned, the other one is guaranteed to be sufficiently large that does not seriously affect the overall performance. When we conduct the tuning experiment on METIS scheme to get the best number of local message process units (n_{lmp}), for example, we manually provide sufficient remote message process units (n_{rmp}) with the DCCM feature off, and then increase n_{lmp} for each PageRank execution instance until the overall performance becomes stable.

Figure 3.6 shows the tuning results of running PageRank on Random and METIS partition schemes, respectively. We do not list results about LDG1 to LDG4, as

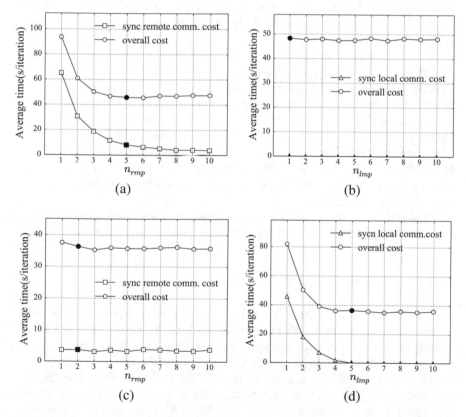

Fig. 3.6 Tuning on Random and METIS partition schemes. Black points are the optimal choices. (**a**) n_{rmp} tuning on Random. (**b**) n_{lmp} tuning on Random. (**c**) n_{rmp} tuning on METIS. (**d**) n_{lmp} tuning on METIS

they all lead to the same conclusions as the above two schemes. The basic rule of determining the proper configurations for a manual tuning is choosing the earliest points where the overall performance is close to the stable one as the best configuration.

As shown in Fig. 3.6a, when the number of remote message process units exceeds five, the overall performance is converged and changes slightly. This indicates the remote message processor with five message process units inside is sufficient for this workload, which is running PageRank on random partitioned uk-2007-05-u. Though, the sync remote communication cost can still decrease a bit with continuously increasing n_{rmp}. But the large number of message process units will also affect other parts of the system (e.g., consuming more computation resources), the overall performance remains stable because of the tradeoff between two factors (i.e., number of message processor units and influence on other parts of the systems).

From Fig. 3.6b, we can easily figure out one local message process unit is sufficient to handle the remained local message processing workload in Random scheme. More message process units do not bring any significant improvement. Based on this tuning experiment, we can see that totally six message process units are enough. Among them, one is for the local message processor, and the other five are for remote message processor.

In Fig. 3.5, the parameters chosen by the DCCM are six message process units for the remote message processor and one for local message processor, when the Random partition scheme is applied. Though they do not exactly match the off-line tuned values, but they fall into the range where the overall performance has been converged and the difference is small. So the DCCM generates a set of parameters almost as good as the best ones acquired from the off-line tuning.

By analyzing Fig. 3.6c, d, we can see seven message process units are sufficient, in which five belong to local message processor and the remained are for remote message processor. This time the DCCM also comes out the similar results as the manually tuned parameters.

Through a bunch of parameter tuning experiments, we verify the effectiveness of the DCCM and find that the DCCM can choose appropriate parameters.

3.4.2.3 Adaptivity of DCCM

In this subsection, we show the results about the adaptivity on various graph partition scheme in PAGE. All the experimental results show that the DCCM is sensitive to both partition quality and initial concurrency setting. It responses fast and can adjust PAGE to a better status within a few iterations.

We run the PageRank on Random and METIS graph partition schemes, with randomly setting the concurrency of the remote message processor and local message processor at the beginning, and let DCCM automatically adjust the concurrency.

Figure 3.7b illustrates each iteration's performance of PageRank running on the METIS partition scheme of uk-2007-05-u. It clearly shows that PAGE can

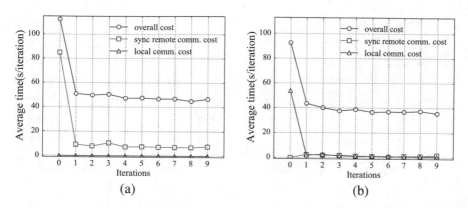

Fig. 3.7 Performance on adaptively tuning by DCCM. (**a**) Random partition scheme. (**b**) METIS partition scheme

rapidly adjust itself to achieve better performance for the task. It costs about 93 s to finish the first iteration, where the sync local communication cost is around 54 s. After first iteration, PAGE reconfigures the concurrency of dual concurrent message processor, and achieves better overall cost in successive iterations by speeding up the process of local messages. The second iteration takes 44 s, and the sync local communication cost is close to zero already. Similar results can be obtained for the Random partitioning scheme, shown in Fig. 3.7a.

3.4.3 Comparison with Other Pregel-Like Systems

In this section, we compare the performance of PAGE with the Pregel-like baselines, i.e., Giraph and Giraph-GPSop. We first present the advantage of PAGE by profiling PageRank execution instance, followed by the evaluation on various graph algorithms. In the end, we show that PAGE can also handle the situation where varying the number of partitions leads to the change of graph partition quality.

3.4.3.1 Advantage of PAGE

We demonstrate that PAGE can maintain high performance along with the various graph partition qualities. Figure 3.8a, b describe the PageRank performance across various partition schemes on PAGE and Giraph. We find that, with the increasing quality of graph partitioning, Giraph suffers from the workload growth of local message processing and the sync local communication cost rises fast. In contrast, PAGE can scalably handle the upsurging workload of local message processing, and maintain the sync local communication cost close to zero. Moreover, the overall performance of PAGE is actually improved along with increasing the quality of

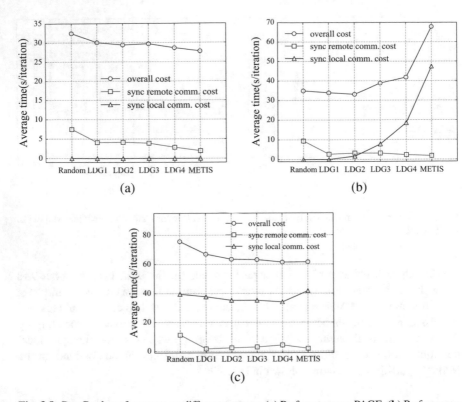

Fig. 3.8 PageRank performance on different systems. (**a**) Performance on PAGE. (**b**) Performance on Giraph. (**c**) Performance on Giraph-GPSop

graph partitioning. In Fig. 3.8a, when the edge cut ratio decreases from 98.52 to 3.48%, the performance is improved by 14% in PAGE. However, in Giraph, the performance is even downgraded about 100% at the same time.

From Fig. 3.8c, we notice the Giraph-GPSop achieves better performance with the improving quality of graph partition as PAGE does. But PAGE is more efficient than Giraph-GPSop over various graph partitions with different qualities. Comparing with Giraph, PAGE always wins for various graph partitions, and the improvement ranges from 10 to 120%. However, Giraph-GPSop only beats Giraph and gains around 10% improvement over METIS partition scheme which produces a really well partitioned graph. For Random partition, Giraph-GPSop is even about 2.2 times worse than Giraph. It is easy to figure out that the central message buffer in Giraph-GPSop leads to this phenomena, as Fig. 3.8c illustrates that the sync local communication cost[1] is around 40 s, though it is stable across six partition schemes.

---

[1]Due to the central message buffer, we treat all the messages as local messages and count its cost into the local communication cost in Giraph-GPSop.

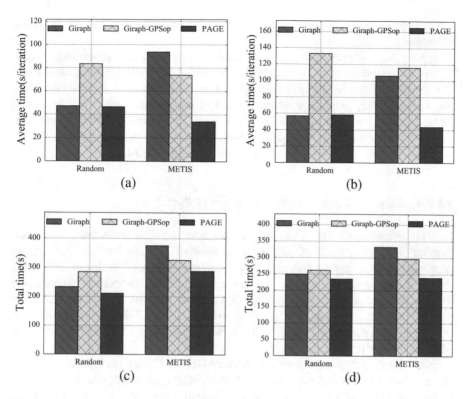

Fig. 3.9 The performance of various graph algorithms. (**a**) PageRank. (**b**) Diameter estimation. (**c**) BFS. (**d**) SSSP

Overall, as a partition-aware system, PAGE can well balance the workloads for local and remote message processing with different graph partitions.

3.4.3.2 Performance on Various Graph Algorithms

In this section, we show that PAGE helps improve the performances of both stationary graph algorithms and non-stationary graph algorithms. The experiments are ran on PAGE, Giraph, and Giraph-GPSop with two stationary graph algorithms and two non-stationary graph algorithms. The two stationary graph algorithms are PageRank and diameter estimation, and the two non-stationary ones are breadth first search (BFS) and single source shortest path (SSSP).

Figure 3.9a, b illustrate the performance of stationary graph algorithms on Random and METIS partition schemes of uk-2007-05-u graph dataset. We find that for the stationary graph algorithms, when the quality of graph partition is improved, PAGE can effectively use the benefit from a high-quality graph partition and improve the overall performance. Compared with the Giraph and Giraph-

GPSop, PAGE outperforms them because PAGE concurrently processes incoming messages in a unified way.

Figure 3.9c, d present the performance of non-stationary graph algorithms. Similar to the stationary graph algorithms, the performance of PAGE surpasses those of Giraph and Giraph-GPSop. This implies PAGE's architecture can also facilitate the non-stationary graph algorithms. However, the performance is not improved when the quality of partition scheme is increased. The reason has been discussed in Sect. 3.2. The non-stationary graph algorithms have a workload imbalance problem, which can be solved by the dynamic partition strategy [11, 24].

3.4.3.3 Performance by Varying Numbers of Partitions

Previous experiments are all conducted on a web graph partitioned into fixed number of subgraphs, i.e., 60 partitions for uk-2007-05-u. In practice, the number of graph partitions can be changed, and different numbers of partitions will result into different partition qualities. We run a series of experiments to demonstrate that PAGE can also efficiently handle this situation. Here we present the results of running PageRank on a social graph, livejournal-u. Table 3.3 lists the edge cut ratios of livejournal-u partitioned into 2, 4, 8, 16, 32, 64 partitions by LDG, Random, and METIS, respectively.

First, Figure 3.10a–c all show that both PAGE and Giraph perform better when the partition number increases across three partition schemes, which is obvious as parallel graph computing systems can benefit more from higher parallelism. When the graph is partitioned into more subgraphs, each subgraph has smaller size, and hence the overall performance will be improved with each worker having less workload. On the other hand, the large number of subgraphs brings heavy communication, so when the partition number reaches a certain threshold (e.g., sixteen in the experiment), the improvement becomes less significant. This phenomenon reveals parallel processing large-scale graph is a good choice, and it will improve the performance.

Second, the improvement between PAGE and Giraph decreases with the increasing number of partitions. As the number of partitions increases, the quality of graph partitioning decreases which means the local message processing workload decreases. Since Giraph performs well over the low quality graph partitioning, the performance gap between PAGE and Giraph is small when the number of partitions is large. Besides, the point where PAGE and Giraph have close performance varies with different graph partitioning algorithms. In METIS scheme, PAGE and Giraph have similar performance around 64 partitions, while in Random scheme, it is about four partitions when they are close. The reason is that different graph partition algorithms produce different partition quality, and the bad algorithms will generate low quality graph partition even the number of partitions is small.

Third, PAGE always performs better than Giraph across three partition schemes for any fixed number of partitions and the reason has been discussed in Sect. 3.4.3.1. But with the increasing number of partitions, the improvement of PAGE decreases.

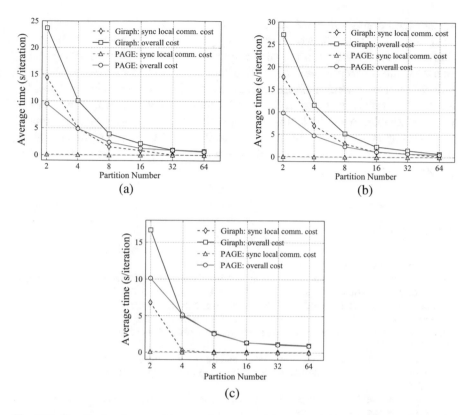

Fig. 3.10 Result of various partition numbers on social graph. (**a**) LDG partition scheme. (**b**) METIS partition scheme. (**c**) Random partition scheme

This is because the workload for each node becomes low when the partition number is large. For a relatively small graph like livejournal, when the partition number is around 64, each subgraph only contains tens of thousand vertices. However, the results are sufficient to show the robustness of PAGE for various graph partitions.

3.5 Summary

In this chapter, we proposed a partition-aware graph computing system PAGE to address the drawbacks in existing systems that they cannot efficiently utilize the high-quality graph partitions. The key technique in PAGE is the dynamic concurrency control model, and the model tracks three high-level critical running metrics and dynamically adjusts the system configurations. In the adjusting model, we introduced two heuristic rules to effectively extract the system characters and generate proper parameters. Furthermore, the proposed technique is a kind of system-level optimization and can be transferred to other distributed systems, like matrix-based system [20], stream computing platform (e.g., Yahoo S4 [18], Twitter Storm [1]), etc. Taking Storm as an example, the basic primitives for doing stream transformations in Storm are "spouts" and "bolts." A spout is a source of streams. A bolt consumes any number of input streams, conducts some processing, and possibly emits new streams. In practice, the parallelism of a blot is specified by user and this is inflexible. Since the logic of a blot satisfies the producer–consumer model, by designing an estimation model that is similar to DCCM, the stream computing platform can also automatically adjust the proper number of processing elements according to the workload.

References

1. *Storm.* http://storm.incubator.apache.org/.
2. Ahmed Amr, Shervashidze Nino, Narayanamurthy Shravan, Josifovski Vanja, and Smola Alexander J. Distributed largescale natural graph factorization. In *WWW*, pages 37–48, 2013.
3. Lars Backstrom, Dan Huttenlocher, Jon Kleinberg, and Xiangyang Lan. Group formation in large social networks: membership, growth, and evolution. In *SIGKDD*, pages 44–54, 2006.
4. P. Boldi and S. Vigna. The webgraph framework i: compression techniques. In *WWW*, pages 595–602, 2004.
5. Paolo Boldi, Marco Rosa, Massimo Santini, and Sebastiano Vigna. Layered label propagation: a multiresolution coordinate-free ordering for compressing social networks. In *WWW*, pages 587–596, 2011.
6. Sergey Brin and Lawrence Page. The anatomy of a large-scale hypertextual web search engine. In *WWW*, pages 107–117, 1998.
7. Avery Ching, Sergey Edunov, Maja Kabiljo, Dionysios Logothetis, and Sambavi Muthukrishnan. One trillion edges: Graph processing at Facebook-scale. *Proc. VLDB Endow.*, 8(12):1804–1815, August 2015.
8. Guojing Cong, George Almasi, and Vijay Saraswat. Fast PGAS implementation of distributed graph algorithms. In *SC*, pages 1–11, 2010.
9. U. Kang, Charalampos E. Tsourakakis, and Christos Faloutsos. Pegasus: A peta-scale graph mining system implementation and observations. In *ICDM*, pages 229–238, 2009.
10. George Karypis and Vipin Kumar. Multilevel algorithms for multi-constraint graph partitioning. In *CDROM*, pages 1–13, 1998.
11. Zuhair Khayyat, Karim Awara, Amani Alonazi, Hani Jamjoom, Dan Williams, and Panos Kalnis. Mizan: a system for dynamic load balancing in large-scale graph processing. In *EuroSys*, pages 169–182, 2013.
12. Jon M. Kleinberg. Authoritative sources in a hyperlinked environment. In *Journal of the ACM*, pages 604–632, 1999.
13. Jure Leskovec, Kevin J. Lang, Anirban Dasgupta, and Michael W. Mahoney. Statistical properties of community structure in large social and information networks. In *WWW*, pages 695–704, 2008.
14. Jure Leskovec, Kevin J. Lang, and Michael Mahoney. Empirical comparison of algorithms for network community detection. In *WWW*, pages 631–640, 2010.
15. Yucheng Low, Danny Bickson, Joseph Gonzalez, Carlos Guestrin, Aapo Kyrola, and Joseph M. Hellerstein. Distributed GraphLab: a framework for machine learning and data mining in the cloud. In *PVLDB*, pages 716–727, 2012.
16. Shuai Ma, Yang Cao, Jinpeng Huai, and Tianyu Wo. Distributed graph pattern matching. In *WWW*, pages 949–958, 2012.

17. Grzegorz Malewicz, Matthew H. Austern, Aart J.C Bik, James C. Dehnert, Ilan Horn, Naty Leiser, and Grzegorz Czajkowski. Pregel: a system for large-scale graph processing. In *SIGMOD*, pages 135–146, 2010.
18. Leonardo Neumeyer, Bruce Robbins, Anish Nair, and Anand Kesari. S4: Distributed stream computing platform. In *ICDMW*, pages 170–177, 2010.
19. John Ousterhout. Why threads are a bad idea (for most purposes). In *USENIX Winter Technical Conference*, 1996.
20. Amitabha Roy, Ivo Mihailovic, and Willy Zwaenepoel. X-stream: Edge-centric graph processing using streaming partitions. In *SOSP*, pages 472–488, 2013.
21. Semih Salihoglu and Jennifer Widom. GPS: a graph processing system. In *SSDBM*, pages 22:1–22:12, 2013.
22. Helmut Simonis and Trijntje Cornelissens. Modelling producer/consumer constraints. In *CP*, pages 449–462, 1995.
23. Isabelle Stanton and Gabriel Kliot. Streaming graph partitioning for large distributed graphs. In *SIGKDD*, pages 1222–1230, 2012.
24. Shang Zechao and Yu Jeffrey Xu. Catch the wind: Graph workload balancing on cloud. In *ICDE*, pages 553–564, 2013.

Chapter 4
Efficient Parallel Subgraph Enumeration

Abstract In this chapter, we introduce a novel parallel subgraph enumeration framework, named PSgL, which is built on top of Pregel-like graph computing systems. The PSgL iteratively enumerates subgraph instances and solves the subgraph enumeration in a divide-and-conquer fashion. The framework completely relies on the graph traversal operation instead of the explicit join operation. To achieve the high efficiency of the framework, we propose several algorithm-specific optimization techniques for balancing the workload and reducing the size of intermediate results. In respect to the workload balance, we theoretically prove the problem of partial subgraph instance distribution is NP-hard, and carefully design heuristic strategies. To reduce the massive intermediate results, we develop three mechanisms, which are automorphism breaking of the pattern graph, initial pattern vertex selection based on a cost model, and a pruning method based on a light-weight index. We implemented the prototype of PSgL, and conducted comprehensive experiments of various graph enumeration operations on real-world large graphs. The experimental results clearly demonstrate that PSgL is robust and can achieve performance gain over the existing considerable solutions up to 90%.

4.1 Introduction

Subgraph enumeration [5] is a basic operation in the frequent subgraph mining, network processing, and motif discovering in bioinformatics [18]. In addition, social network analyses also heavily rely on subgraph enumeration to reveal the information cascade patterns in real life [15]. This operation aims to find all the occurrences of a pattern graph in a data graph. Figure 4.1 lists a square pattern graph G_p, and a data graph G_d. The operation is to find all the subgraph instances in G_d that are isomorphic with G_p. In this example, the found subgraph instances are $\square_{1235}, \square_{1256}, \square_{2345}$.

Though it is very useful, the subgraph enumeration is computationally challenging [1]. The size of the result set is often exponential to the number of vertices in the pattern graph. Most of the algorithms enumerate the subgraph instances one by one [5, 10], and cannot handle large graphs. Stream-based approaches [4, 23] can

© Springer Nature Singapore Pte Ltd. 2020
Y. Shao et al., *Large-scale Graph Analysis: System, Algorithm and Optimization*,
Big Data Management, https://doi.org/10.1007/978-981-15-3928-2_4

Fig. 4.1 A toy example of subgraph enumeration. (**a**) Pattern graph G_p. (**b**) Data graph G_d

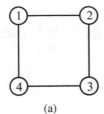

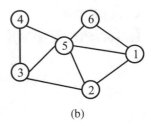

(a)   (b)

process the large graphs, but they only output the approximate occurrence number, and the isomorphic subgraph instances are not available. Later on, MapReduce-based parallel solutions [1, 19] were proposed for the subgraph enumeration. The approach [1] follows the parallel query processing paradigm. It decomposes the pattern graph into small ones (edges), extracts the subgraphs for each small pattern graph, and joins the intermediate results finally. The performance of the parallel solution is mainly restricted by the expensive join operator. Besides, the imbalance distribution of data graph or intermediate results also hinders the performance of these solutions.

Here we introduce a novel parallel subgraph enumeration framework, named PSgL. PSgL is originated from the fact that each subgraph instance is independent. More concretely, the subgraph instances can be enumerated concurrently without confliction. Therefore, in the view of the subgraph instances, the subgraph enumeration operation is an embarrassingly parallel problem [7]. We can execute the operation in a parallel and divide-and-conquer fashion, to improve the performance of subgraph enumeration in large graphs. In PSgL, the subgraph enumeration operation is iteratively divided into partial subgraph enumerations by the graph traversal, and each worker expands the *partial subgraph instance*, meanwhile, communicates with each other through the distribution strategy. The process is proceeded until all the subgraph instances are found. The whole execution is based on the graph traversal, thus avoiding the costly join operation. Recall the example in Fig. 4.1, the existing approaches [1, 19] will generate a path of length four through join operation and verify the equality of two ends for the square pattern graph. However, in PSgL, after creating the path of length three by graph traversal, it can verify the partial subgraph instance fast through checking the neighborhood of one end without generating paths of length four. This implies that PSgL will not produce more complicated intermediate results with the help of graph traversal.

However, there still are two challenges to efficiently perform subgraph enumeration in PSgL. One is the workload balance problem. The imbalance is the curse for the performance improvement to a parallel computing framework. In every iteration of PSgL, a totally different set of partial subgraph instances is processed, i.e., the workload characteristic varies sharply between iterations. The other is the enormous partial subgraph instance problem. Since the result set exponentially increases when the size of the pattern graph grows, the partial subgraph instances will be even more than the result set due to the invalid ones. These enormous valid and invalid partial subgraph instances heavily affect the computation, communication, and memory cost.

To address the first challenge, we prove the hardness of the workload balance problem in PSgL, that is NP-hard. We resort to heuristics and carefully design three distribution strategies for the PSgL when it distributes the partial subgraph instances in each iteration. They are random distribution strategy, roulette wheel distribution strategy, and workload-aware distribution strategy. For the workload-aware distribution strategy, we design a cost model for partial subgraph instance expansion to estimate workload, and then introduce a simple but effective heuristic rule to achieve a satisfied workload balance.

For the second one, we first break the automorphism of the pattern graph, thus guaranteeing each subgraph instance is found exactly once and reducing the number of duplicated partial subgraph instances during the runtime. We next present another cost model to select a good initial pattern vertex which has the least number of intermediate partial subgraph instances. Third, we design a light-weight edge index to effectively filter the invalid partial subgraph instances as early as possible.

We implemented the prototype of PSgL, and run comprehensive experiments on large-scale graphs. The results demonstrate that PSgL outperforms the previous considerable parallel solutions.

The rest of this chapter is organized as follows: we introduce the preliminaries in Sect. 4.2. In Sect. 4.3, we elaborate the PSgL framework followed by presenting our optimization techniques in Sect. 4.4. We show the experimental results in Sect. 4.5 and conclude the work in Sect. 4.6.

4.2 Problem Definition and Backgrounds

4.2.1 Problem and Notations

Subgraph enumeration is a classic operation in graph computing. It enumerates all the instances of a pattern graph in a data graph, and both graphs have no labels on vertices and edges. We design the PSgL framework to efficiently execute the operation on undirected graphs *in parallel*. A graph is denoted by $G = (V, E)$, where V and E are the sets of vertices and edges. For each vertex $v \in V$, $N(v)$ denotes the neighborhood of v, and $deg(v)$ is the degree of v, which equals to $|N(v)|$. We distinguish the pattern graph and data graph by subscripts p and d.

Now we proceed to present several key conceptions for the analysis of PSgL.

Power-Law Graph is a graph whose degree distribution follows a power law. That is, the probability of a vertex having a degree d is given by

$$p(d) \propto d^{-\gamma},$$

where the parameter γ is a positive constant that controls the skewness of the degree distribution. A lower γ indicates that more vertices are high degree, i.e., more skewed.

Random Graph is a graph that is generated by some random processes. The classic random model is the Erdős–Rényi model [6]. The degree distribution of the ER random graph follows the Poisson distribution, and most of the vertices have the degrees around the average.

Ordered Graph is an undirected graph with manual assignment of partial order for the vertices. The data graph G_d is ordered by following two rules:

1. for any $v_d, u_d \in V_d$, if $deg(v_d) < deg(u_d)$, then $v_d < u_d$;
2. if $deg(v_d) = deg(u_d)$ and v_d has a smaller vertex id, then $v_d < u_d$.

For a vertex v_d in the ordered graph, we use n_b to denote the number of neighbors who have smaller rank, and n_s to represent the counterpart. The property of n_b and n_s is described below.

Property 4.1 The distribution of n_b is more skewed than the original degree distribution, while n_s is more balanced.

Here we give a brief explanation. Assume in the original data graph, the probability that a vertex has a degree d is $p(d)$. Then, for a vertex v_d, whose degree is d, n_s and n_b are

$$n_b = d \times \sum_{d_i < d} p(d_i), \quad n_s = d \times \left(1 - \sum_{d_i < d} p(d_i) \right) \qquad (4.1)$$

From Eq. 4.1, it is easy to figure out that higher d derives higher n_b and lower d leads to lower n_b. This implies the distribution of n_b will be more polarized compared with the original. In contrast, the distribution of n_s will be more concentrated to the average and more balanced. Taking a power-law graph, WebGoogle, as an example, the original degree distribution has $\gamma = 1.66$. After ordering it, γ is 1.54 for the n_b distribution while it has $\gamma = 3.97$ for the n_s distribution.

4.2.2 Partial Subgraph Instance

Partial subgraph instance is a data structure that records the mapping between G_p and G_d, denoted by G_{psi}. G_{psi} consists of $|V_p|$ vertex mapping pairs, where each pair is represented by $<v_p, v_d>$ ($v_d = map(v_p)$). Thus, assume the vertices of G_p are numbered from 1 to $|V_p|$, we can also simply state G_{psi} as $\{map(1), map(2), \ldots, map(|V_p|)\}$. In Fig. 4.1, for example, the G_{psi} for the original G_p is $\{?, ?, ?, ?\}$, where "?" means the v_p has no mapped v_d, while the G_{psi} for $\square_{1256}$ is $\{1, 2, 5, 6\}$. In addition, we use *subgraph instance* to represent the found subgraph in data

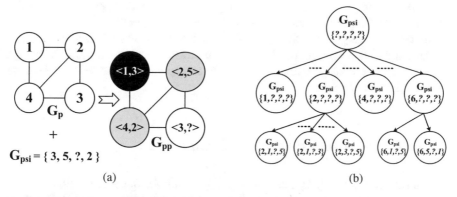

Fig. 4.2 Partial pattern graph and partial subgraph instance tree. (a) G_{pp}. (b) G_{psi} tree

graph, such as $\square_{1235}$, $\square_{1256}$, $\square_{2345}$. The original pattern graph combining a partial subgraph instance forms a *partial pattern graph*, G_{pp}, as illustrated in Fig. 4.2a. The vertex color of G_{pp} will be explained in Sect. 4.3.3.

4.2.3   Automorphism of a Pattern Graph

The automorphism of a pattern graph $G_p = (V_p, E_p)$ is a permutation σ of the vertex set V_p, such that the pair of vertices (u_p, v_p) forms an edge if and only if the pair $(\sigma(u_p), \sigma(v_p))$ also forms an edge. That is, it is a graph isomorphism from G_p to itself. The automorphism causes the same subgraph instance found multiple times. For example, without any preprocessing, the $\square_{2345}$ can be found eight times by the pattern graph in Fig. 4.1, because there exist eight valid permutations that make the square pattern graph isomorphism to itself. In order to guarantee that each subgraph instance is found exactly once, we need to eliminate the automorphism of a pattern graph. We name this procedure as *automorphism breaking*, and the approach is described in Sect. 4.4.2.1.

4.3   Parallel Subgraph Enumeration Framework

The parallel subgraph enumeration framework, named PSgL, follows the Pregel-like graph computing paradigm [17], which applies the vertex-centric programming model and bulk synchronous parallel [21]. The data graph is randomly distributed among the workers' memory. PSgL iteratively expands the partial subgraph instances by data vertices in parallel, until all the subgraph instances are found. In each iteration, the newly created partial subgraph instances are sent across the workers according to the distribution strategy. The entire enumeration process relies on the partial subgraph instance and its independence property.

In the following subsections, we discuss the properties of partial subgraph instance first. Then we elaborate the PSgL framework based on the partial subgraph instance and the core partial subgraph instance expansion algorithm. At last, we analyze the cost of PSgL.

4.3.1 Independence Property of Enumeration

Partial subgraph instance is a key intermediate result for the enumeration process. All the partial subgraph instances during an enumeration form a tree, called G_{psi} tree. A node in the tree corresponds to a G_{psi} and the children of a node are derived from expanding one mapped data vertex in the node (Sect. 4.3.3). The root of the tree is the G_{psi} obtained from the input pattern graph. Figure 4.2b illustrates (a part of) the G_{psi} tree for square pattern graph enumeration on data graph in Fig. 4.1. $\{?,?,?,?\}$ is the root. By expanding the mapped data vertex 6 in $\{6,?,?,?\}$, it generates two new G_{psi}s, $\{6,1,?,5\}$ and $\{6,5,?,1\}$. Therefore, $\{6,1,?,5\}$ and $\{6,5,?,1\}$ are the children of $\{6,?,?,?\}$ in the tree.

A single G_{psi} encodes a set of subgraph instances in its subtree. G_{psi}s at the same level of the tree are a complete representation of the whole result set. From top to down in a G_{psi} tree, the original enumeration task is divided into more fine-grained subtasks (subtrees). Moreover, we notice that the partial subgraph instance is independent from each other except the ones in its generation path. In other words, after the partial subgraph instance is generated, it can be processed independently without being aware of other ones. For example, $\{2,1,?,3\}$ and $\{6,5,?,1\}$ can be simultaneously computed on data vertex 1 and 5, respectively. We name this property as the *independence property*.

The tree hierarchy and independence property allow the subgraph enumeration problem to be solved in a **divide-and-conquer** style. We can first divide the problem into partial subgraph enumeration and then conquer the partial subgraph instance and generate new ones in parallel, the process is repeated until all results are returned.

4.3.2 PSgL Framework

Our proposed PSgL is in charge of constructing the G_{psi} tree for a pattern graph in parallel. It consists of two distinct phases, i.e., initialization and expansion, and both are concentrated on the partial subgraph instance.

Initialization Phase In this phase, each data vertex creates a G_{psi} which only contains a vertex mapping pair of the data vertex and the selected initial pattern

---

Algorithm 1 Expand a partial subgraph instance

Input: partial subgraph instance G_{psi}, pattern graph G_p
 1: get the current expanding v_p and v_d from G_{psi}
 2: $candList \leftarrow NULL, color(v_p) \leftarrow$ BLACK
 3: /* explore v_p's neighbor */
 4: **foreach** v'_p in $N(v_p)$ **do**
 5:     **if** processNeighbor(v'_p,v_d,$candList$)=**false then**
 6:         **return**;
 7:     **end if**
 8: **end foreach**
 9: **if** isComplete(G_{psi})=**true then**
10:     **print** subgraph instance
11: **else**
12:     generate new G_{psi}s by combining candidates in $candList$ and
        distribute them based on distribution strategy in Algorithm 3.
13: **end if**

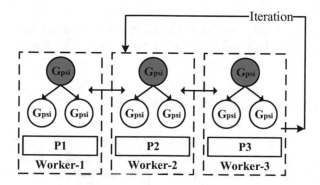

Fig. 4.3 Iterative process in PSgL framework

vertex (Sect. 4.4.2.2). These G_{psi}s form the initial set of partial subgraph instances and correspond to a one-level G_{psi} tree as the start.

Expansion Phase This is the core procedure in the PSgL framework and the whole G_{psi} tree will be constructed after this phase. Figure 4.3 shows the high-level execution flow of expansion phase. It consists of several iterations. In an iteration, where the divide-and-conquer happens, each G_{psi} is expanded independently on a certain worker, and generates more fine-grained ones. These new G_{psi}s are sent to the next worker according to the distribution strategy, if they are not the complete subgraph instances. The iteration ends when the previously received G_{psi}s are processed and the new ones are received by the corresponding workers. The phase will not finish until there is no G_{psi} in the framework.

Please note that PSgL may not guarantee that each G_{psi} is expanded in the same pace, which means that the G_{psi} tree does not grow level by level. It only guarantees

---

Algorithm 2 Process a neighbor of pattern vertex

Input: v_p', v_d, $candList$
Output: updated $candList$
  1: **if** color(v_p') = GRAY **then**
  2:     **if** the $map(v_p')$ is not in $N(v_d)$ **then**
  3:         **return false**
  4:     **end if**
  5: **else if** color(v_p') = WHITE **then**
  6:     /* details are described in Algorithm 5*/
  7:     $candidate$ ← getCandidateSet(v_p')
  8:     **if** $candidate$ = NULL **then**
  9:         **return false**
 10:     **end if**
 11:     add the $candidate$ into $candList$
 12: **end if**{BLACK v_p' is omitted.}
 13: **return true**

---

that all the G_{psi}s in PSgL cover the results which have not been found. Nevertheless, PSgL is also suitable for the case where the tree grows one level in each iteration.

4.3.3 Partial Subgraph Instance Expansion

The partial subgraph instance expansion algorithm generates more fine-grained G_{psi}s and makes the G_{psi} tree grow. It explores the partial pattern graph, which contains three kinds of vertices: BLACK, GRAY, and WHITE. Figure 4.2a shows a G_{pp} during the expansion.

BLACK vertex is the one which has been expanded. The neighbors of a BLACK vertex must consist of BLACKs and GRAYs.

GRAY vertex has a mapped data vertex, but it has not been expanded. It must be adjacent to at least one BLACK vertex. Moreover, the GRAY vertices represent the expanding candidate set of this partial subgraph instance for the following process.

WHITE vertex is the one which has not been mapped to any data vertex.

The expansion algorithm changes one GRAY vertex, which is specified by the previous iteration, into BLACK, and also makes its neighbor become GRAY if it is WHITE. The expansion procedure is presented in Algorithm 1. It first obtains the expanding pattern vertex v_p from G_{psi}, sets v_p in BLACK (Lines 1–2), and then processes v_p through neighbor exploration on the pattern graph (Lines 4–8). In the end (Lines 9–13), it prints the found subgraph instance, if the G_{psi} is completed. Otherwise, it sets the WHITE neighbors in GRAY, and creates new G_{psi}s by combining neighbors' candidates with pruning invalid combinations via the rules in Sect. 4.4.2.3. For each new G_{psi}, it selects a GRAY vertex according to the distribution strategy as the next expanding vertex and sends the new G_{psi} to the corresponding worker.

Algorithm 2 illustrates the details for processing a neighbor of v_p. It processes the neighbor according to its color. If it is GRAY, we need to verify whether the edge exists in the data graph or not. It retrieves the candidates for the WHITE vertex from the neighbors of data vertex v_d. BLACK vertices are just omitted because they have been expanded previously and the corresponding edge must exist in the data graph.

4.3.4 Cost Analysis

In this subsection, we analyze the performance of PSgL and reveal the metrics related to its cost. Because the partial subgraph instance is the minimal computing unit in the framework, let us first consider the cost of processing a single partial subgraph instance in the expansion phase, denoted as $load(G_{psi})$. From Algorithm 1, the $load(G_{psi})$ consists of communication and computation costs. The communication cost is caused by distributing the new G_{psi}s, while the generating process itself causes the significant computation cost. The asynchronous communication model [17] facilitates PSgL to do the communication and computation concurrently, so the $load(G_{psi})$ is determined by the heavier one. On account of the problem itself is computation intensive, the $load(G_{psi})$ is approximate to the computation cost, in most iterations.

The computation cost includes the cost of verifying the GRAY neighbors ($cost_g$) and the cost of generating new G_{psi}s by retrieving candidates for the WHITE neighbors ($cost_w$). We use c_e to represent the cost of generating a single G_{psi} and $f(v_p)$ to denote the number of new G_{psi}s through expanding v_p. The $load(G_{psi})$ can be calculated as

$$load(G_{psi}) = cost_g + cost_w = cost_g + c_e \times f(v_p). \qquad (4.2)$$

Based on the $load(G_{psi})$, the cost of a worker can be represented as $L = \sum_{i=1}^{N} load(G_{psi})$, where N is the number of G_{psi} processed by the worker.

Assume that the subgraph enumeration task runs on a system with K workers, and the task finishes in S iterations. During the ith iteration, the worker k processes N_{ki} G_{psi}s. Then we have the total cost of the task T as

$$T = \sum_{i=1}^{S} \max_{1 \leq k \leq K} \{L_{ki}\} = \sum_{i=1}^{S} \max_{1 \leq k \leq K} \left\{ \sum_{j=1}^{N_{ki}} load(G_{psi})_j \right\} \qquad (4.3)$$

To achieve high performance in PSgL, it is required to minimize the total cost T. From the above equation, we can figure out there are three metrics affecting the overall cost.

First is the number of iterations S. In general, the fewer number of iterations achieves better performance, since there is some overhead between iterations, like synchronization. The following theorem shows the bound of S for a pattern graph.

Theorem 4.1 *Given a pattern graph $G_p = (V_p, E_p)$, if the G_{psi} tree grows level by level, then S is restricted by $|MVC| \leq S \leq |V_p| - 1$, where $|MVC|$ is the size of the minimum vertex cover of G_p.*

Proof Because the G_{psi} tree grows level by level, in each iteration, only one v_p is expanded. Thus PSgL must visit at least $|MVC|$ vertices before every edge in G_p is explored. But it will not exceed $|V_p| - 1$ after all the edges are explored. In conclusion, S is limited by $|MVC| \leq S \leq |V_p| - 1$. $\square$

In PSgL, not all the partial subgraph instances have the same expansion path because of the distribution strategy, and hence we do not explicitly optimize the S. Its effects are considered during the initial pattern vertex selection.

The second issue is the workload balance which is implied by the max function. As a classic parallel computing framework, it can be damaged by the imbalance. So we design and compare several distribution strategies in Sect. 4.4.1 to investigate which one provides better performance.

The last is the number of G_{psi}s. As above equations indicate that the performance of worker in an iteration is related to N_{ki} and $f(v_p)$, we need the expansion algorithm produce as small size of G_{psi} as it can. This will reduce the computation cost and communication cost at the same time. We propose several techniques to achieve this goal in Sect. 4.4.2.

4.4   The Optimizations of the Framework

In this section, we introduce the optimization techniques to improve the performance of PSgL. These techniques are classified into two categories. One is the partial subgraph instance distribution strategy for the workload balance, and the other is for reducing the enormous partial subgraph instances.

4.4.1   Workload Distribution Strategy

As Sect. 4.3.4 discussed, the distribution of G_{psi}s should make the workload balance among the workers in an iteration. In general, a good data graph partition will help to achieve this goal. However, due to the generality of the input pattern graph and the variable characteristic of workload of G_{psi}s between iterations, it is difficult to design a one-size-fit-all graph partition.

In PSgL, the data graph is simply random partitioned, and the G_{psi}s are distributed online without fixed expansion path to avoid the workload imbalance. For a single G_{psi}, there are several choices when distributing it. Illustrated in Fig. 4.2a, all the GRAY vertices are the candidates for the next expansion. It is possible to achieve workload balance by selecting a good destination for each G_{psi}.

The cost of a G_{psi} has been modeled in Eq. 4.2. Here we define the increased workload w_i of worker i, if a G_{psi} is sent to the worker i. The w_i can be formulated as

$$w_i = \begin{cases} \min_{\text{GRAY } v_p} \{load(G_{psi})_{v_p}\} & map(v_p) \text{ belongs to } i; \\ +\infty & \text{otherwise.} \end{cases} \quad (4.4)$$

We proceed to define the **partial subgraph instance distribution problem** as below:

Definition 4.1 There are N partial subgraph instances to be processed by K workers, and the cost of a partial subgraph instance processed by worker i is defined in Eq. 4.4. We use W_i to denote the total cost for the partial subgraph instances processed by worker i. Then the goal is to find a distribution strategy for the N partial subgraph instances to achieve

$$\min \left\{ \max_{1 \le i \le K} \{W_i\} \right\}$$

The following theorem describes the hardness of the partial subgraph instance distribution problem.

Theorem 4.2 *The partial subgraph instance distribution problem is NP-hard.*

Proof Minimum Makespan Scheduling [13], which is a well-known NP-hard problem, can be easily reduced to the partial subgraph instance distribution problem. A typical Minimum Makespan Scheduling problem on unrelated machines can be stated as "There are m parallel machines and n independent jobs. Each job is to be assigned to one of the machines. The processing of job j on machine i requires time P_{ij}. The objective is to find a schedule that minimizes the makespan." So if jth job maps to the jth G_{psi} and the cost of job j processed by machine i maps to w_{jk}, then the Minimum Makespan Scheduling problem is reduced to the partial subgraph instance distribution problem. □

In order to solve this NP-hard problem efficiently, we propose the following heuristic rules.

4.4.1.1 Workload-Aware Distribution Strategy

For the partial subgraph instance distribution problem, all the partial subgraph instances are generated online, so we need an online solution. The classical heuristic rule for this situation is selecting worker j for the ith G_{psi} which has minimal overall workload,

$$\arg \min_{j} \{W_j + w_{ij}\}$$

In [13], the authors proved that this greedy rule is tightly bounded by $K \times OPT$. However, this greedy solution is likely to obtain a local optimum as it always tries to balance the cost first when a new item comes in.

We refine the heuristic rule by reducing the penalty part and it will increase the opportunity to jump out the local optimum. The general form of the heuristic rule can be

$$\arg\min_{j}\{W_j^\alpha + w_{ij}\}, 0 \le \alpha \le 1,$$

where W_j^α stands for the penalty part which restricts the objective to be balanced. When $\alpha = 1$, it is the original heuristic rule. When $\alpha = 0$, it implies that each time the rule chooses worker j where the G_{psi} incurs the least increased workload. But this rule is more likely to be imbalanced, though the total workload would be minimized.

Based on the above observations, we choose $\alpha = 0.5$ and make a tradeoff between the balance and minimal workload. To some extent, the rule with $\alpha = 0.5$ can avoid local optimum more likely than $\alpha = 1$, while it achieves more balance than $\alpha = 0$. Besides, the following theorem guarantees that in the worst case, the performance of $\alpha = 0.5$ is still bounded by $K \times OPT$.

Theorem 4.3 *For the partial subgraph instance distribution problem, the overall cost achieved by the heuristic rule with $\alpha = 0.5$ is no K times worse than the OPT.*

Proof Let $s_i = \min_{1 \le j \le K} w_{ij}$ which indicates the minimal cost of ith G_{psi} processed across all the K workers and each w_{ij} is positive integer in our problem.

Then let the lower bound of the cost for n G_{psi} as

$$g(n) = \sum_{i=1}^{n} s_i$$

It is easy to infer that $OPT \ge \frac{1}{K} \times g(N)$. Let $f(n)$ represent the overall cost obtained by the greedy algorithm. Each time when ith G_{psi} is distributed to worker k, the following inequations are held:

$$W_k^\alpha + w_{ik} \le W_j^\alpha + w_{ij}, 1 \le j \le K \tag{4.5}$$

We next prove the theorem by inducing the task id.

(1) $i = 1$, $f(1) = s_1 \le g(1)$
(2) $f(i) = \max(f(i-1), W_k + w_{ik})$

   (2a) $f(i-1) \ge W_k + w_{ik}$, then $f(i) = f(i-1) \le g(i-1) \le g(i-1) + s_i \le g(i)$

Algorithm 3 Distribution strategy summary

1:   ***Roulette Wheel Distribution Strategy***
2: **foreach** GRAY v_p in G_{psi} **do**
3:     $p_{v_p} \leftarrow$ calculate through Eq. 4.6.
4: **end foreach**
5: $randnum \leftarrow$ Random(0,1)
6: **foreach** GRAY v_p in G_{psi} **do**
7:     **if** $randnum \leq p_{v_p}$ **then**
8:         **return** v_p;
9:     **end if**
10:     $randnum \leftarrow randnum - p_{v_p}$
11: **end foreach**

1:   ***Workload Aware Distribution Strategy***
2: W_j indicates the total workload for worker j
3: $i \leftarrow$ current G_{psi} id
4: $k \leftarrow 0$ // k records the selected worker id
5: v_k records the selected expanding v_p
6: w_{ik} records the increased workload for worker k
7: **foreach** GRAY v_p in G_{psi} **do**
8:     $j \leftarrow$ worker id of $map(v_p)$
9:     **if** $W_k^\alpha + w_{ik} > W_j^\alpha + \left(\frac{deg(v_d)}{w_{v_p}}\right)$ **then**
10:         $k \leftarrow j, w_{ik} \leftarrow \left(\frac{deg(v_d)}{w_{v_p}}\right), v_k \leftarrow v_p$
11:     **end if**
12: **end foreach**
13: $W_k \leftarrow W_k + w_{ik}$
14: **return** v_k

(2b) $f(i-1) \leq W_k + w_{ik}$, then $f(i) = W_k + w_{ik}$, and Eq. 4.5 makes sure that $W_k^{0.5} + w_{ik} \leq W_j^{0.5} + s_i \leq f(i-1)^{0.5} + s_i$.

First, we can get the inequations: $f(i) - w_{ik} \leq f(i-1) \leq f(i)$

Then,

$$W_k^{0.5} + w_{ik} \leq f(i-1)^{0.5} + s_i$$

$$[f(i) - w_{ik}]^{0.5} + w_{ik} \leq f(i-1)^{0.5} + s_i$$

$$(w_{ik} - s_i)^2 \leq f(i-1) + [f(i) - w_{ik}]$$

$$- 2f(i-1)^{0.5} \times [f(i) - w_{ik}]^{0.5}$$

$$\leq f(i-1) + [f(i) - w_{ik}]$$

$$- 2[f(i) - w_{ik}]^{0.5} \times [f(i) - w_{ik}]^{0.5}$$

$$f(i) \leq f(i-1) + s_i + (w_{ik} - s_i) - (w_{ik} - s_i)^2$$

$$\leq f(i-1) + s_i \leq g(i-1) + s_i \leq g(i)$$

At last, we obtain $f(N) \leq g(N) \leq K \times OPT$.   □

In the following, we show how to estimate $load(G_{psi})$ in practice. Because $cost_g$ is only caused by neighborhood search, which can be done efficiently by a bitmap index, we can estimate the cost of a G_{psi} as $f(v)$. The value of $f(v_p)$ is limited to $[0, \binom{deg(v_d)}{w_{v_p}}]$ range, here w_{v_p} is the number of WHITE neighbors around v_p. Subsequently, we use the upper bound of $f(v)$ to estimate itself, thus guaranteeing the estimated value and the accurate $f(v)$ have the same order. So $load(G_{psi}) \simeq \binom{deg(v_d)}{w_{v_p}}$, if it expands v_p for the G_{psi}. The pseudocode of the workload-aware distribution strategy is presented in Algorithm 3. The time complexity of the algorithm can be $O(|V_p|)$ by only calculating the approximate value of $\binom{deg(v_d)}{w_{v_p}}$.

4.4.1.2 Naive Distribution Strategies

Here we introduce two other distribution strategies for the partial subgraph instance distribution problem. They are random distribution strategy and roulette wheel distribution strategy.

Random distribution strategy is randomly choosing a GRAY vertex for a G_{psi}. It will balance the number of G_{psi} processed by a worker. However, the different G_{psi}s will have various effects to the workload and the cost is still imbalance among workers. Clearly, this strategy has its own merit, i.e., it is simple and has the minimal overhead.

Roulette wheel distribution strategy considers the degree information of the data graph when distributing a partial subgraph instance. Since $load(G_{psi})$ is approximated to $f(v_p)$, it implies the larger degree of a data vertex results in higher cost for a G_{psi}. Thus the following heuristic rule can be derived.

Heuristic The data vertex with larger degree should expand less G_{psi}.   □

The rule indicates G_{psi} is preferred to be expanded by a data vertex with small degree. Considering the balance problem, we propose a distribution strategy based on the roulette wheel selection [2].

The roulette wheel distribution strategy chooses GRAY vertex k in G_{psi} with a probability p_k. p_k is defined in the following equation:

$$p_k = \frac{\prod_{j=1, j \neq k}^{n} deg(v_{dj})}{\sum_{i=1}^{n} \prod_{j=1, j \neq i}^{n} deg(v_{dj})}, \tag{4.6}$$

where n is the number of GRAYs in G_{psi}, v_{dj} is the data vertex mapping to GRAY vertex v_{pj}.

The probability indicates that G_{psi} has a higher chance to be expanded by a data vertex with smaller degree. In order to avoid the imbalance, it is still possible that some G_{psi}s are distributed to the high degree data vertices. The strategy can achieve better balance than the random strategy. However, due to the probability

p_k is unaware of previous G_{psi}'s distribution, it may cause some data vertices with small degree overloaded.

As the probability of each v_p can be calculated in $O(1)$ with proper preprocess, the roulette wheel distribution is a linear strategy with time complexity $O(|V_p|)$. The procedure is described in Algorithm 3.

4.4.2 Partial Subgraph Instance Reduction

The enormous partial subgraph instances consume a lot of memory resources, and introduce expensive generation and communication cost as well. It is important to reduce the number of partial subgraph instances, in order to improve PSgL's performance. However, the size of partial subgraph instances is related to many factors, including the structure of the pattern graph, structure of the data graph, the initial pattern vertex, distribution strategies, and so on. It is tough to design one solution for such a complex problem. We propose three independent mechanisms from three aspects to reduce the size of partial subgraph instances. They are automorphism breaking of the pattern graph, initial pattern vertex selection based on a cost model, and a pruning method based on a light-weight index.

4.4.2.1 Automorphism Breaking of the Pattern Graph

As mentioned in Sect. 4.2, automorphism breaking of a pattern graph guarantees each subgraph instance is found exactly once, it reduces the duplicated partial subgraph instances. However, the classical graph labeling method [8] for breaking automorphism cannot handle our problem, because the labeling has nothing to do with the data graph, which even has no label.

Since the data graph has been ordered by its degree sequence, we also assign a partial order set for the pattern graph. The partial order set not only breaks the automorphism of the pattern graph, but also can be used to prune partial subgraph instances. Because the graph automorphism problem is still unknown whether it has a polynomial time algorithm or it is NPC problem [16], breaking it has the same difficulty. Here we introduce an approach by iteratively assigning a partial order on the pattern graph until the graph is not automorphism anymore.

The main idea, in each iteration, is to pick an equivalent vertex group, where each vertex can be mapped to others in a certain automorphism of the pattern graph, and eliminate a member from the group by assigning a partial order, which sets the lowest rank to the eliminated member compared to the remained ones. We use the DFS to find the equivalent vertex group in a pattern graph, and it has been demonstrated that DFS can detect automorphism of a graph with up to 100 vertices in seconds [10].

Algorithm 4 Cost estimation for a pattern vertex

Input: v_p, G_p
1: $estimatedCost \leftarrow 0$
2: $l \leftarrow 0, n \leftarrow |V_d|, G_{pp} \leftarrow G_p$ /*l means the number of iteration.*/
3: mark v_p GRAY in G_{pp}.
4: $queue \leftarrow (G_{pp}, n, l)$
5: **while** $queue$ is not empty **do**
6:     $(G_{pp}, n, l) \leftarrow queue.front()$; $queue.pop()$
7:     $estimatedCost \leftarrow estimatedCost + cost(G_{pp}, n, l)$
8:     **if** G_{pp} is expandable **then**
9:         **foreach** $GRAY\ v_p$ in G_{pp} **do**
10:            expand and generate new $(G'_{pp}, n', l + 1)$
11:            **if** $(G'_{pp}, n', l + 1)$ exists **then**
12:                update the existed $(G'_{pp}, n', l + 1)$
13:            **else**
14:                $queue.push((G'_{pp}, n', l + 1))$
15:            **end if**
16:        **end foreach**
17:    **end if**
18: **end while**
19: **return** $estimatedCost$

However, there are several partial order sets to break the automorphism of a graph. Because PSgL follows the graph traversal, the partial order on edges can be immediately used during the exploration, and prune invalid partial subgraph instances early. Moreover, as the order between vertices who are not connected can only be used after exploration, we apply the following heuristic rule to select a good partial order set.

Heuristic In each iteration, the algorithm selects the equivalent vertex group which contains vertices with higher degree to break.   □

4.4.2.2   Initial Pattern Vertex Selection

Initial pattern vertex is the one where the algorithm starts to traverse. Fixing an initial pattern vertex prevents producing duplicated subgraph instances. However, different initial pattern vertices generate partial subgraph instances in different size and have various influences on the performance of a pattern graph. We design a cost model-based initial pattern vertex selection for the general pattern graph. Besides, we derive a deterministic rule based on the model for two special pattern graphs: cycles and cliques.

General Pattern Graph   The optimal initial pattern vertex should be the one which leads to the minimal cost. We design a cost model to estimate the cost of a certain initial pattern vertex, and select the vertex with minimal estimated cost as the "best" initial pattern vertex.

For a general pattern graph, the initial pattern vertex selection framework enumerates all the pattern vertices, and for each one, calculates its cost by traversing from the selected vertex along with estimating the cost for each generated partial pattern graph (Algorithm 4). The overall estimated cost for an initial pattern vertex equals to the sum of the cost of all the partial pattern graphs. The time complexity of selection framework is $O(|V_p| \times |E_p|)$.

The core portion of the framework is the model to estimate the cost of a partial pattern graph, $cost(G_{pp}, n, l)$. In the selection framework, we assume that the random distribution strategy is used. When there are n G_{psi}s for G_{pp} to be expanded, the cost of these G_{psi}s can be estimated as

$$cost(G_{pp}) = n \times \left(cost_g + \frac{1}{C} \sum_{i=1}^{C} c_e \times f(v_{pi}) \right),$$

where C is the number of GRAY vertices in G_{pp}, and $cost_g$ is similar for different GRAY vertices.

As it does not know the data vertex that v_p maps to, the approach in Sect. 4.4.1 fails to estimate $f(v_p)$. But, it is easy to obtain the degree distribution $p(d)$ of the data graph by sampling or traversing, we can estimate $f(v_p)$ with the following equation:

$$f(v_p) \simeq \sum_{d=deg(v_p)}^{d_{max}} p(d) \times \binom{d}{w_{v_p}}$$

Based on the selection framework and cost estimation model, we can choose a good initial pattern vertex for the general pattern graph.

Cycles and Cliques The initial pattern vertex selection framework indicates the following theorem.

Theorem 4.4 *Under the initial pattern vertex selection framework, the best initial pattern vertex is the one which derives a traversal order minimizing the total number of partial subgraph instances.*

Proof Assuming the simulation for a pattern vertex terminates in S iterations and, in the lth iteration, there exist T_l different partial pattern graphs, each has n_{lt} corresponding G_{psi}s. The total number of G_{psi} in iteration l is n_l. As the random distribution strategy is applied, so $n_l = n_{lt} \times T_l$. Then the total estimated cost for the pattern vertex is

$$T_e = \sum_{l=0}^{S} \sum_{t=1}^{T_l} n_{lt} \times \left(cost_g + \frac{1}{C} \sum_{i=1}^{C} c_e \times f(v_{pi}) \right)$$

Next, we define g_l as the *average expanding coefficient* at iteration l. The g_l can be represented as

$$g_l = \begin{cases} 1 & l = 0 \\ \frac{1}{T_l} \frac{1}{C} \sum_{t=1}^{T_l} \sum_{i=1}^{C} f(v_{pi}) & l \neq 0. \end{cases}$$

Now, T_e can be represented by g_l

$$T_e = \sum_{l=0}^{S} (n_l \times cost_g + c_e \times n_l \times g_l) \propto \sum_{l=0}^{S} n_l \times g_l \tag{4.7}$$

Here we ignore $cost_g$ with the same reason in Sect. 4.4.1.

The term $n_l \times g_l$ is the number of newly generated G_{psi} in iteration l. So from Eq. 4.7, we can conclude, under our selection framework, a traversal order minimizing the total number of partial subgraph instances results into minimal estimated cost T_e, which implies the corresponding initial pattern vertex is the best. $\square$

For the cycles and cliques, after breaking the automorphism, there must be a vertex who has the lowest rank, because the first equivalent vertex group contains all the pattern vertexes. Then we have a ***deterministic rule***, that is the vertex with the lowest rank is the best initial pattern vertex for the cycles and cliques, and the following theorem shows the correctness.

Theorem 4.5 *After breaking the automorphism of cycles and cliques, the vertex v_{lr} with the lowest rank is the best initial pattern vertex for any ordered data graph.*

Proof

(I) First step, $l = 1$.

$$g_1 = f(v_p) = \frac{1}{|V_d|} \sum_{i=1}^{|V_d|} \binom{deg(v_{di})}{w_{v_p}} \quad \because C = 1, T_1 = 1$$

For cliques and cycles, $w_{v_p} = |V_p| - 1$ and $w_{v_p} = 2$, respectively. On an ordered graph with the partial order pruning, $deg(v_{di}) = n_s$ or n_b. Because $\sum_{i=1}^{|V_d|} n_s (or\ n_b) = |E_d|$, and Property 4.1 holds, we can easily infer that v_{lr} has the minimal g_1.

(II) Remained steps, $l > 1$. For all the cliques we have $g_l = 1$, because of $w_{v_p} = 0$. For the cycles, due to $w_{v_p} = 1$, all the g_ls are linear to the degree and result in $g_i \simeq g_j, j > i > 1$ (*sophisticated analyses are omitted*).

At last, Eq. 4.7 can derive

$$T_e \propto \sum_{l=0}^{S} n_l \times g_l = n_0 \times \sum_{l=0}^{S} \prod_{i=0}^{i=l} g_i \qquad (4.8)$$

Based on the characteristics of g_l and Eq. 4.8, we can work out that v_{lr}, which has minimal g_1, leads to the minimal number of partial subgraph instances, for the cycles and cliques. According to Theorem 4.4, v_{lr} is the best initial pattern vertex.

□

Though, Theorem 4.5 points out v_{lr} is the best initial pattern vertex, the improvement still depends on the original degree distribution of the data graph. For the power-law graph, where the distributions of n_b and n_s can be totally different,[1] v_{lr} can enhance the performance significantly. Given a random graph, where the n_b and n_s are similar after ordering, PSgL may not benefit a lot from v_{lr}.

4.4.2.3 Pruning Invalid Partial Subgraph Instance

The aforementioned two techniques reduce the size of partial subgraph instances off-line. However, during the runtime, many invalid partial subgraph instances, which do not lead to find other subgraph instances, are generated. The later they are pruned, the more resources they consume.

In order to reduce the number of invalid partial subgraph instances, the quality of the candidate set of WHITE vertices needs to be improved. This can be done by efficient filtering rules. However, without label information, most existing pruning techniques [11, 24] fail in this context. The only information we can use is the graph structure and the partial order. First, the degree information filters the partial subgraph instance, if $deg(v_d) < deg(v_p)$. Second, it is the neighbor connectivity. When retrieving candidates for a WHITE neighbor of the expanding pattern vertex v_p, it should guarantee that the edge between the neighbor and v_p's GRAY neighbor exists in the data graph. In Fig. 4.2a, we need to check edge (3,4) when expanding vertex 2. At last, during the expansion, the partial order must be consistent between the pattern graph and data graph.

Since the data graph is stored in distributed memory, it is expensive to check an edge's existence remotely. So we design a light-weight edge index in PSgL for fast checking the existence of an edge in data graph. It is an inexact index which is built on the bloom filter [3], and indexes the ends of an edge. The index can be built in $O(m)$ time and consumes a small memory footprint. Moreover, the precision of the index is adjustable and the successive iteration only needs to verify a small portion of partial subgraph instances.

---

[1] Refer to the example in Sect. 4.2.

Algorithm 5 Get candidate set

Input: a WHITE neighbor v'_p of v_p
Output: candidates *cand*
 1: **foreach** v'_d in $N(v_d)$ **do**
 2:     /*pruning rule 1.*/
 3:     **if** $deg(v'_d) < deg(v'_p)$ **and** partial order restriction **then**
 4:         **continue**
 5:     **end if**
 6:     /*pruning rule 2.*/
 7:     *valid* ← **true**
 8:     **foreach** v''_p in $N(v_p)$ **do**
 9:         **if** $color(v''_p) =$ GRAY **and** checkEdgeExistence(v'_d, map(v''_p)) = **false then**
10:             *valid* ← **false**
11:             **break**
12:         **end if**
13:     **end foreach**
14:     **if** $valid =$ **false then**
15:         **continue**
16:     **end if**
17:     add u'_d into *cand*
18: **end foreach**

With the help of above pruning rules, it can generate high-quality candidate sets. Algorithm 5 illustrates the procedure of candidate generation for a WHITE vertex. It first uses the degree constrains and partial order restriction to filter the invalid candidates, and then checks all the v_p's GRAY neighbors through the light-weight index.

4.4.3 Implementation Details

In this section, we present the implementation details of PSgL. The prototype of PSgL is written in Java on Giraph,[2] which is an open-source Pregel first released by Facebook.

The design of initialization phase and expansion phase in PSgL follows the vertex-centric model, so both phases are integrated into a single vertex program on Giraph. The first superstep (iteration) of the vertex program is responsible to execute the initialization phase. In the following supersteps, each data vertex processes the incoming G_{psi}s by Algorithm 1. The messages communicated among workers not only include G_{psi}, but also encode the status information, such as the next expanding pattern vertex, the colors of pattern vertices, and the progress of G_{psi}. This is the basic implementation of PSgL, and the G_{psi} tree grows one level in each iteration. Besides the basic vertex program, PSgL also requires several kinds of

---

[2]https://github.com/apache/giraph.

shared data, i.e., pattern graph, initial pattern vertex, light-weight index, and degree statistics. Considering these data that are all small enough to be stored on a single node, (i.e., the edge index of Twitter dataset only costs 2 GB), in current version of PSgL, each worker maintains a copy of them. Moreover, these shared data are static, so we compute them off-line once and load them before running the vertex program through preApplication() API in WorkerContext object.

Furthermore, the distributor is a specific module of PSgL for supporting various distribution strategies. Since the distributor selects the next expanding pattern vertex for a G_{psi} and is shared by all the local data vertices, it can be initialized in preApplication() as well. The two naive distribution strategies only rely on the static shared information, which are easy to implement locally. The workload-aware distribution strategy ideally needs the dynamical global information of each worker's workload W_i. However, in the parallel execution setting, it is expensive to maintain such a global view. Instead, during the distribution, each worker only maintains a local view of the entire workload distribution. Thus the update of W_k can be done fast without communication and synchronization. In practice, since a partition usually contains a moderate size of vertices, it is possible to make a good distributing decision according to the local information.

4.5 Experiments

In this section, we evaluate the performance of PSgL. The following subsection describes the environment, datasets, and pattern graphs. We experimentally demonstrate the effectiveness of optimization techniques in Sect. 4.5.2 and 4.5.3. In Sect. 4.5.4, we present the comparison results on various pattern graphs. At last, we evaluate the scalability of PSgL on large graphs and the number of workers in Sect. 4.5.5.

4.5.1 Experimental Setup

All the experiments were conducted on a cluster with 28 nodes, where each node is equipped with an AMD Opteron 4180 2.6 GHz CPU, 48 GB memory, and a 10 TB disk RAID.

Pattern Graphs We use five different pattern graphs, PG_1–PG_5, illustrated in Fig. 4.4. The partial orders obtained from automorphism breaking are listed below the pattern graph.

Graph Datasets We use six real-world graphs and one synthetic graph in our experiments. All the real-world graphs are undirected ones created from the original release by adding reciprocal edge and eliminating loops and isolated nodes. Except Wikipedia, which can be downloaded from KONECT, other real-world graphs are

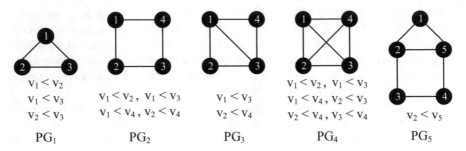

$$v_1 < v_2$$
$$v_1 < v_3 \qquad v_1 < v_2,\; v_1 < v_3 \qquad v_1 < v_3 \qquad v_1 < v_2,\; v_1 < v_3$$
$$v_2 < v_3 \qquad v_1 < v_4,\; v_2 < v_4 \qquad v_2 < v_4 \qquad v_1 < v_4,\; v_2 < v_3 \qquad v_2 < v_5$$
$$\qquad\qquad\qquad\qquad\qquad\qquad\qquad\qquad\qquad v_2 < v_4,\; v_3 < v_4$$

$$PG_1 \qquad\qquad PG_2 \qquad\qquad PG_3 \qquad\qquad PG_4 \qquad\qquad PG_5$$

Fig. 4.4 Pattern graphs

Table 4.1 Meta-data of graphs

	WebGoogle	WikiTalk	UsPatent	LiveJournal	Wikipedia	Twitter	RandGraph
$|V|$	0.9M	2.4M	3.8M	4.8M	26M	42M	4M
$|E|$	8.6M	9.3M	33M	85M	543M	1202M	80M

available on SNAP. The random graph is generated by NetworkX following the Erdős–Rényi model. Table 4.1 summarizes the meta-data of these graphs.

During the experiments, we only output the occurrences of the pattern graph. But we generate the found subgraph instances and can store them if necessary. Each experiment is ran five times and we showed the average performance without including the loading time. The reported runtime is the real job execution time which includes both communication and computation costs.

4.5.2   Effects of Workload Distribution Strategies

We evaluate five distribution variants, i.e., random distribution strategy, roulette wheel distribution strategy, and workload-aware distribution strategy with three different parameters, which are named (WA,0), (WA,0.5), (WA,1). Take (WA,0) as an example, it means the workload-aware distribution strategy which has $\alpha = 0$.

Since the distribution strategy balances the workload for each iteration, the effect of distribution strategies varies according to the different characteristics of an iteration. Here we show the results of PG_2, where middle ($l > 1$) iterations generate new partial subgraph instances, and PG_4, where only the first ($l = 1$) iteration generates partial subgraph instances, as representatives. We profiled the experiments and verified the effectiveness of the (WA,0.5) strategy.

From Table 4.2, we can see that the strategy (WA,0.5) can achieve about 77% improvement on the WikiTalk against the random distribution strategy, when running PG_2. Compared with other three strategies, it can still have around 11% to 23% improvement. There are similar results on the WebGoogle. On the USPatent, the improvement is not as significant as previous two graphs. This is because the

Table 4.2 Runtime (seconds) of various distribution strategies

Data graph	PG	Random	Roulette	(WA,1)	(WA,0)	(WA,0.5)
WebGoogle	PG_2	116.036	69.260	60.256	66.996	50.141
WikiTalk	PG_2	18.285	5.573	5.078	4.853	4.272
UsPatent	PG_2	73.987	75.164	74.008	72.742	62.052
LiveJournal	PG_4	12.957	13.147	12.762	12.897	12.856

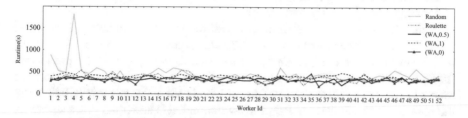

Fig. 4.5 Each worker's performance on WikiTalk with PG_2

degree distributions of WebGoogle and WikiTalk are seriously power-law skewed, which have $\gamma = 1.66$ and $\gamma = 1.09$, respectively, while the power-law parameter γ of USPatent is 3.13. It reveals that the strategy (WA,0.5) is obviously beneficial for the graphs with skewed degree distribution, when the pattern graph generates new partial subgraph instances in the middle iteration.

Furthermore, all the five strategies have similar efficiency when running PG_4 on the LiveJournal. For the clique pattern graph, it only generates the partial subgraph instances in the first iteration which is affected by the initial pattern vertex, and the following iterations are for the verification, which is an operation with constant cost. So the other distribution strategies can also obtain good performance. This implies the performance of a distribution strategy is related to the pattern graph as well.

Figure 4.5 shows the detailed performance of each worker when running PG_2 on WikiTalk. It illustrates that the strategy (WA,0.5) achieves balance while minimizing the cost of the slowest worker. Though, the strategy (WA,1.0) achieves similar balance, it is stuck into the local optimum and cannot minimize the cost of the slowest worker. While strategy (WA,0) can guarantee most workers have smaller cost, but it cannot achieve better balance, the worker 35 performs much slower than the others. In addition, the slowest worker is different between random distribution strategy and roulette wheel distribution strategy. It is because the vertices with higher degree cause the imbalance in random distribution strategy, while the ones with smaller degree having too much workload cause the imbalance in roulette wheel distribution strategy. These phenomena are consistent with previous discussions in Sect. 4.4.1.

4.5.3   Effects of Partial Subgraph Instance Reduction

4.5.3.1   Importance of the Initial Pattern Vertex

Here we report the results on the power-law graph and random graph with running PG_1, PG_2, and PG_4, which have a deterministic rule to identify the good initial pattern vertex according to the cost model. This will clearly reason the importance of selecting a good initial pattern vertex.

Figure 4.6a–c show the performance of different initial pattern vertices on the power-law graph. The real runtime of each initial pattern vertex is normalized to the runtime of the best initial pattern vertex for each pattern graph, so we present the runtime ratio in figures. For the clarity of the figure, we did not present the runtime ratio which exceeds 100 times over the best initial pattern vertex on WikiTalk. From the figures, we notice that for PG_1 on LiveJournal, it is about 8.5 times slower by selecting the highest rank v_3 as the initial pattern vertex than selecting the lowest rank v_1. Note that the performance of v_2 is similar to the one of v_1, because there exists an edge (v_2, v_3) with order $<$ for the v_2 in PG_1. The gap is even larger on WikiTalk, which reaches about 285 times. The clique pattern graph PG_4 has the similar results. The gap is 4 times on LiveJournal, while it is 106.4 times on WikiTalk. On the web graph, WebGoogle, the initial pattern vertex of PG_1 and PG_4, has the similar effects as on the social graph. The improvements are 8.4 and 14.6 times, for PG_1 and PG_4, respectively. Therefore it is necessary to choose a

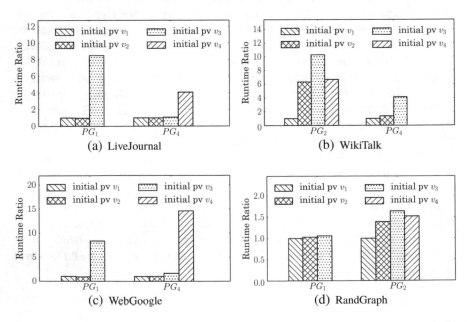

Fig. 4.6 Influences of the initial pattern vertex on various data graphs. (**a**) LiveJournal. (**b**) WikiTalk. (**c**) WebGoogle. (**d**) RandGraph

Table 4.3 Pruning ratio of the edge index

Data graph	PG	$\#G_{psi}$ w/ index	$\#G_{psi}$ w/o index	Pruning ratio
LiveJournal	$PG_1(v_1)$	2.86×10^8	6.81×10^8	58.01%
	$PG_4(v_1)$	9.93×10^9	OOM	Unknown
USPatent	$PG_5(v_1)$	2.26×10^7	3.17×10^8	92.87%
	$PG_5(v_3, v_4)$	7.38×10^9	2.04×10^{10}	63.89%

$PG_i(v_j)$ stands for the number of G_{psi} counted during the expansion of v_j for the PG_i

good initial pattern vertex when enumerating a certain pattern graph on real-world graphs. The deterministic rule in Theorem 4.5 is effective for this task.

However, Fig. 4.6d shows that all the three initial pattern vertices in PG_1 have the similar performance on the random graph, and for PG_2, the gap is only about 1.6 times between the slowest and fastest initial pattern vertexes. It indicates that the influence of initial pattern vertex is less significant on random graph than the one on power-law graph, for the cycles and cliques. This is consistent with previous analysis in Sect. 4.4.2.2.

4.5.3.2 Efficiency of the Light-Weight Edge Index

Next we present the results on the efficiency of the light-weight edge index. With the help of the light-weight edge index, PSgL can filter invalid partial subgraph instances early, and saves the communication and memory costs.

Table 4.3 only lists the online statistics for executing the pattern graph PG_1, PG_4, PG_5 on LiveJournal and USPatent, as representatives due to the space constraint, as the other pattern graphs give similar results. We notice that the pruning ratio can be as high as 92.87% during the expansion of v_1 for the PG_5 on USPatent. Without the edge index, the 92.87% invalid partial subgraph instances will cause heavy communication and memory consumption. On LiveJournal, when running PG_4 starting from v_1 without the edge index, the task fails with Out-Of-Memory[3] (OOM for short) problem because of the enormous invalid partial subgraph instances. Furthermore, in the different iterations, the index has different pruning ratio. It has small pruning ratio in the later iterations during which the size of partial subgraph instances is closer to the result set.

In summary, if there exists the invalid partial subgraph instances, the edge index can prune them efficiently and reduce the memory and communication overhead.

---

[3]Using the terminology in Java to denote the error.

4.5.4 Performance on Various Pattern Graphs

Now we evaluate the performance of PSgL on the real datasets with various pattern graphs, PG_1–PG_5. All the optimization techniques discussed in Sect. 4.4 are used. We compare PSgL with the MapReduce solutions (i.e., Afrati [1] and SGIA-MR [19]). For one input pair of pattern graph and data graph, the cost of each solution is normalized to the cost of PSgL. The normalized cost is called runtime ratio and is presented in figures. A runtime ratio x for a solution $\mathcal{A}$ means the performance of $\mathcal{A}$ is x times slower than PSgL's performance. For the clarity of the figures, the ratios exceed 100 times are not visualized. In addition, as the MapReduce solutions cannot be finished in 4 h for PG_5, we did not show the results in figures either, and the results of PG_3 on LiveJournal are omit for the same reason.

Figure 4.7 lists the runtime ratio between three solutions. We can easily see from the figure, that PSgL significantly outperforms the MapReduce solutions on the WikiTalk, WebGoogle, and USPatent. On average, PSgL can achieve performance gains over the MapReduce solutions around 90%. Especially, when executing PG_4 on USPatent, the runtime ratio between PSgL and Afrati can be 225 (not visualized). The join operation makes the reducer operate slowly. Besides the graph traversal advantage in PSgL, the online distribution strategy helps PSgL avoid the serious imbalance and achieves a good performance across various settings. In contrast, the MapReduce solutions have varieties of performance across the different datasets,

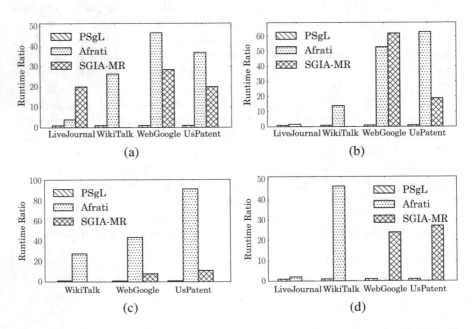

Fig. 4.7 Runtime ratio among PSgL, Afrati, and SGIA-MR. (**a**) PG_1. (**b**) PG_2. (**c**) PG_3. (**d**) PG_4

and the two surpass each other interleaved. This is because the distribution strategy of intermediate results in these solutions follows a fixed scheme and the degree of the skewness for each MapReduce solution changes sharply with different graphs. For example, SGIA-MR spends 4213 s for PG_1 on WikiTalk and Afrati finishes in only 190 s. However, on WebGoogle, Afrati is about 1.6 times slower than SGIA-MR.

Furthermore, though, the speed up of PSgL on LiveJournal is around 2–4 times across different pattern graphs, the absolute saved time is significant. For example, when running PG_2, PSgL finishes in 4302 s, while Afrati consumes 7291 s. For PG_5, even though, we assume the MapReduce solutions finish in 4 h, the speed up of PSgL is around 3–14 times on different data graphs.

4.5.5 Scalability

4.5.5.1 Scalability on Large Graphs

To evaluate the scalability of our approach on large graphs, we further conduct experiments on two large datasets, Twitter and Wikipedia, and also compare with GraphChi [14] and PowerGraph [9]. GraphChi and PowerGraph are two state-of-the-art graph computing systems on a single node and in parallel, respectively. For both methods, the latest C++ versions are used in the experiments. Meanwhile, we show the robustness of our proposal by comparing with a one-hop index based solution on PowerGraph.

Table 4.4 shows the comparison results of triangle counting on Wikipedia and Twitter, since triangle (PG_1) is a very typical pattern graph and attracts more attention in social network analysis. We can see that PSgL achieves performance gains over the MapReduce solution up to 97% on Twitter and 86% on Wikipedia. It surpasses GraphChi as well. However, compared with PowerGraph, PSgL is about 4–6 times slower. One reason is that PowerGraph is a heavily optimized graph-parallel execution engine [22]. Furthermore, the triangle counting operation on PowerGraph is well optimized via a one-hop neighborhood index maintained by hopscotch hashing [12].

By using the one-hop index technique, it is insufficient to enable PowerGraph to solve a general pattern graph efficiently. We extend the graph traversal based solution in PSgL to the PowerGraph. Unlike the original PSgL, we manually choose a traversal order for the general pattern graph, and let PowerGraph solve

Table 4.4 Triangle listing on large graphs

Data graph	Pattern graph	Afrati	PowerGraph(C++)	GraphChi(C++)	PSgL
Twitter	PG_1	432 min[a]	2 min	54 min	12.5 min
Wikipedia	PG_1	871 s	36 s	861 s	125 s

[a]The performance is cited from [20]

Table 4.5 General pattern graph listing comparison

Data graph	Pattern graph	Traversal order	Afrati	PowerGraph	PSgL
WikiTalk	PG_2	1->2->3->4	4402 s	48 s	318 s
WikiTalk	PG_3	2->3->4->1	13,743 s	100 s	494 s
WikiTalk	PG_3	1->2->3->4	13,743 s	OOM	494 s
WikiTalk	PG_4	1->2->3->4	1785 s	127 s	38 s
LiveJournal	PG_4	1->2->3->4	2749 s	OOM	1330 s
WebGoogle	PG_5	1->2->3->4->5	>4 h	OOM	4232 s

the subgraph listing based on that order and prune the intermediate results via the one-hop index. The traversal order is denoted by like "A->B->C," which means the algorithm first visits vertex "A," then "B" and "C" on the pattern graph. Furthermore, we eliminate the automorphism of the pattern graph to guarantee each result is found once as well.

Table 4.5 illustrates the performances of PSgL and PowerGraph on general pattern graphs. For the simple pattern graph PG_2, similar to the triangle PG_1, PowerGraph can obtain better performance with the help of one-hop index. While pattern graphs become complicated, PowerGraph degrades because of the enormous invalid intermediate results, and cannot handle general pattern graphs even the data graph is small. For example, when executing PG_4 on WikiTalk, PowerGraph is 3.3 times slower than PSgL. The result is even worse on LiveJournal, the task is failed with the Out-Of-Memory (OOM) problem. This is because, without the global edge index, the algorithm is unable to use the connectivity except the one-hop link to prune the invalid intermediate results, thus burdening the memory and communication overhead. Moreover, unlike PSgL where the distribution strategy dynamically chooses the traversal order for each G_{psi}, the fixed traversal order cannot achieve a balanced distribution of intermediate results. When running PG_5 on PowerGraph, the imbalanced distribution leads to OOM on some nodes. Another important observation is, similar to the initial pattern vertex selection in PSgL, the different fixed traversal orders heavily affect the performance and it is difficult for a non-expert to figure out a good traversal order for the general pattern graphs. For instance, when executing PG_3 with traversal order "2->3->4->1," it leads to better performance while the performance degrades significantly with the traversal order "1->2->3->4." Because "1->2->3->4" makes the algorithm generate a large set of initial intermediate results and causes the OOM issue.

In summary, PSgL is a subgraph listing framework to support the general pattern graphs by addressing above problems via the workload-aware distribution strategy, a cost model-based initial pattern vertex selection method and the light-weight edge index. The experimental results show PSgL's scalability and robustness with respect to various pattern graph types and datasets.

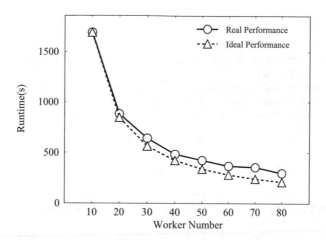

Fig. 4.8 Performance vs. worker number

4.5.5.2   Scalability on the Number of Workers

Finally we demonstrate that PSgL has a graceful scalability with the increasing number of workers. Figure 4.8 shows the performance of running PG_2 on the WikiTalk with the number of workers raising from 10 to 80. We notice that the real performance curve is approximate to the ideal curve, which assumes the performance is linear to the worker number. The runtime is decreased closely linear with respect to the worker number. For example, it is about 1691 s for running PG_2 with 10 workers, while the cost is reduced to 845 s when the workers are doubled. As the number of workers goes up, the improvement decreases slightly.

4.6   Summary

In this work, we have proposed an efficient parallel solution, called PSgL, to address the subgraph enumeration problem on large-scale graphs. PSgL is a parallel iterative subgraph listing framework, which is graph friendly and designed based on the basic graph operation. Moreover, we introduced several optimization techniques to balance the workload and reduce the size of intermediate results, which can further enhance the performance of PSgL. Through the comprehensive experimental study, we demonstrated that PSgL outperforms the state-of-the-art solutions in general.

References

1. Foto N. Afrati, Dimitris Fotakis, and Jeffrey D. Ullman. Enumerating subgraph instances using map-reduce. ICDE, 2013.
2. Thomas Bäck. Evolutionary algorithms in theory and practice: evolution strategies, evolutionary programming, genetic algorithms. Oxford University Press, 1996.
3. Burton H. Bloom. Space/time trade-offs in hash coding with allowable errors. Commun. ACM, 1970.
4. I. Bordino, D. Donato, A. Gionis, and S. Leonardi. Mining large networks with subgraph counting. ICDM, 2008.
5. Norishige Chiba and Takao Nishizeki. Arboricity and subgraph listing algorithms. SIAM J. Comput., 1985.
6. P. Erdős and A. Rényi. On random graphs. I. Publ. Math. Debrecen, 1959.
7. Ian Foster. Designing and building parallel programs: Concepts and tools for parallel software engineering. AWL Co., Inc., 1995.
8. Joseph A. Gallian. A dynamic survey of graph labeling. Electron. J. Combin., 2000.
9. Joseph E. Gonzalez, Yucheng Low, Haijie Gu, Danny Bickson, and Carlos Guestrin. Power-Graph: distributed graph-parallel computation on natural graphs. In *OSDI*, 2012.
10. Joshua A. Grochow and Manolis Kellis. Network motif discovery using subgraph enumeration and symmetry-breaking. RECOMB, 2007.
11. Huahai He and Ambuj K. Singh. Graphs-at-a-time: query language and access methods for graph databases. SIGMOD, 2008.
12. Maurice Herlihy, Nir Shavit, and Moran Tzafrir. Hopscotch hashing. DISC, 2008.
13. Oscar H. Ibarra and Chul E. Kim. Heuristic algorithms for scheduling independent tasks on nonidentical processors. J. ACM, 1977.
14. Aapo Kyrola, Guy Blelloch, and Carlos Guestrin. GraphChi: Large-scale graph computation on just a pc. OSDI, 2012.
15. Jure Leskovec, Ajit Singh, and Jon Kleinberg. Patterns of influence in a recommendation network. PAKDD, 2006.
16. Anna Lubiw. Some NP-complete problems similar to graph isomorphism. SIAM J. Comput., 1981.
17. Grzegorz Malewicz, Matthew H. Austern, Aart J.C Bik, James C. Dehnert, Ilan Horn, Naty Leiser, and Grzegorz Czajkowski. Pregel: a system for large-scale graph processing. SIGMOD, 2010.
18. R. Milo, S. Shen-Orr, S. Itzkovitz, N. Kashtan, D. Chklovskii, and U. Alon. Network motifs: simple building blocks of complex networks. Science, 2002.
19. Todd Plantenga. Inexact subgraph isomorphism in MapReduce. J. Parallel Distrib. Comput., 2013.
20. Siddharth Suri and Sergei Vassilvitskii. Counting triangles and the curse of the last reducer. WWW, 2011.
21. Leslie G. Valiant. A bridging model for parallel computation. Commun. ACM, 1990.
22. Reynold S. Xin, Joseph E. Gonzalez, Michael J. Franklin, and Ion Stoica. GraphX: a resilient distributed graph system on spark. GRADES Workshop In SIGMOD, 2013.
23. Zhao Zhao, Maleq Khan, V. S. Anil Kumar, and Madhav V. Marathe. Subgraph enumeration in large social contact networks using parallel color coding and streaming. ICPP, 2010.
24. Gaoping Zhu, Xuemin Lin, Ke Zhu, Wenjie Zhang, and Jeffrey Xu Yu. TreeSpan: efficiently computing similarity all-matching. SIGMOD, 2012.

Chapter 5
Efficient Parallel Graph Extraction

Abstract In this chapter, we introduce the homogeneous graph extraction task, which extracts homogeneous graphs from the heterogeneous graphs. In an extracted homogeneous graph, the relation is defined by a line pattern on the heterogeneous graph and the new attribute values of the relation are calculated by user-defined aggregate functions. When facing large-scale heterogeneous graphs, the key challenges of the extraction problem are how to efficiently enumerate paths matched by the line pattern and aggregate values for each pair of vertices from the matched paths. To address the above two challenges, we propose a parallel graph extraction framework. The framework compiles the line pattern into a path concatenation plan, which is selected by a cost model. To guarantee the performance of computing aggregate functions, we first classify the aggregate functions into distributive aggregation, algebraic aggregation, and holistic aggregation; then we speed up the distributive and algebraic aggregations by computing partial aggregate values during the path enumeration. The experimental results demonstrate the effectiveness of the proposed graph extraction.

5.1 Introduction

The real world is fruitful. There are many interesting objects in reality, such as persons, groups, and locations. The relations among objects have diverse semantics as well. To model these real-world multi-typed relations, heterogeneous graph has been widely used. In Fig. 5.1, we list a toy example of heterogeneous scholarly graph which depicts three types of relations among three types of entities (i.e., authors, papers, and venues) in a research community. There are authorBy (dashed-arrow) relations between the Author vertices and Paper vertices; citeBy (dotted-arrow) relations are between Paper vertices; and publishAt (solid-arrow) relations connect the Paper and Venue vertices. Analyzing heterogeneous graphs with these abundant semantics is an important approach to find interpretive results and meaningful conclusions [18].

Most of existing graph analysis algorithms, like SimRank [15, 21], community detection [14, 16], centrality computation [9], etc., focus on such homogeneous

© Springer Nature Singapore Pte Ltd. 2020
Y. Shao et al., *Large-scale Graph Analysis: System, Algorithm and Optimization*,
Big Data Management, https://doi.org/10.1007/978-981-15-3928-2_5

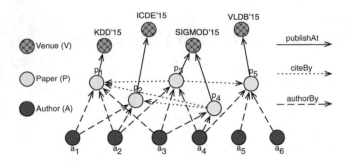

Fig. 5.1 A tiny scholarly graph. There are three types of vertices, Author, Paper, and Venue, and three types of relations, publishAt, authorBy, and citeBy

graphs. Directly executing them on heterogeneous graphs by simply ignoring the semantics of vertices and edges, the results will lose their values. Consequently, graph extraction is introduced as a basic preprocessing step for analyzing the heterogeneous graphs. Graph extraction outputs a subgraph only containing single-typed edges, thus eliminating the gap between heterogeneous graphs and classic graph analysis solutions. We call the output subgraph as *edge homogeneous graph*. Furthermore, graph extraction can also facilitate the new heterogeneous graph analysis approaches to process large graphs. Many proposed heterogeneous graph analysis solutions [18] select proper features (e.g., linear combination of different relations [1], meta-path [6, 19]) by extracting new relations from the original heterogeneous graphs.

Before introducing the challenges of extracting edge homogeneous graph extraction over large graphs, we present some concepts of the extraction via an example. In Fig. 5.2, we list four possible edge homogeneous graphs extracted from the aforementioned heterogeneous scholarly graph. Figure 5.2a is a co-author graph, in which the co-author relation is defined as that "two Author vertices have authorBy relations to a same Paper vertex on the scholarly graph." The pattern (right-above figure in Fig. 5.2a) to define the new relation is called *line pattern*. To find all the new relations in the scholarly graph, we need to enumerate paths satisfying the line pattern between vertices. This process is called *path enumeration*. The attribute values of new relation are computed with the user-defined aggregate functions. In the co-author graph example, the aggregate functions simply count the number of paths between a pair of vertices. The value of edge (a_3, a_4) is two, which means there are two different paths satisfying the line pattern. Given that the aggregation is computed between every pair of vertices, we name it as *pair-wise aggregation* in the graph extraction context. The formal definition of the homogeneous graph extraction problem is presented in Sect. 5.2.

However, existing techniques cannot efficiently solve the extraction problem on large heterogeneous graphs. Regular path query (RPQ) [10] is a good option for the path enumeration. But RPQ requires the linear number of iterations with regard to the length of line pattern and incurs massive intermediate results. Another possible

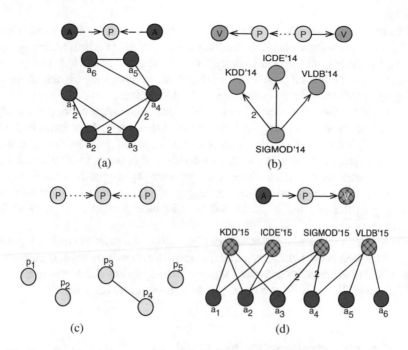

Fig. 5.2 Four edge homogeneous graph extraction instances on the scholarly graph. In each subfigure, the line pattern is above its extracted homogeneous graph. The attribute value is the number of paths between each pair of vertices, and the value is not shown when it is one. (**a**) Co-author graph. (**b**) co-citation venue graph. (**c**) Co-citation paper graph. (**d**) Author-venue graph

solution is to use graph databases. But the graph databases are only optimized for querying local structures, and cannot efficiently handle the extraction problem neither. Furthermore, the computation of pair-wise aggregation is another challenge. The original evaluation of RPQ does not involve aggregations. Since there are massive paths during the extraction, it is expensive if we simply compute the aggregation path by path.

We introduce a parallel graph extraction framework over large heterogeneous graphs. The basic idea is that long paths can be generated by concatenating short paths, so we compile the input line pattern into a path concatenation plan. The plan contains a set of primitive patterns, which are line patterns of length two, and the plan also determines the order of concatenating paths. Since the primitive pattern has length two, the two end vertices are both the neighbors of the middle vertex. In accordance with such property, we use the vertex-centric model [7] to iteratively evaluate the path concatenation plan in parallel. After the paths are generated, the pair-wise aggregation can also be computed in the vertex-centric model naturally.

In general, a line pattern corresponds to several path concatenation plans (Fig. 5.5), and different plans lead to different performance. We design a cost model to guide the framework find a high-performance path concatenation plan. On the basis of the cost model, we find that both the number of iterations and the number

of intermediate paths heavily influence the performance of the framework. Consequently, we design three plan selection strategies—iteration optimized strategy, path
optimized strategy, and hybrid strategy. The best strategy is the hybrid one which
generates the minimized number of intermediate paths among all the plans with
$O(log(l))$ iterations, where l is the length of a line pattern.

To improve the efficiency of computing pair-wise aggregation, we analyze
the properties of different aggregate functions and classify them into distributive
aggregation, holistic aggregation, and algebraic aggregation. The distributive and
algebraic aggregations can be solved in a divide-and-conquer fashion which helps
reduce the number of intermediate paths by merging the paths before they are
completely enumerated, while holistic aggregation can only be computed in a
path-by-path manner and sophisticated techniques are required to achieve high
performance.

We organize the remaining as follows: we define the problem of homogeneous
graph extraction in Sect. 5.2. The parallel graph extraction framework is presented in
Sect. 5.3. The aggregation and plan selection techniques are elaborated in Sects. 5.4
and 5.5, respectively. In Sect. 5.6, we present the experimental results.

5.2 Graph Extraction Problem

5.2.1 Preliminaries

A heterogeneous graph is an abstraction of multi-typed relations among various
objects in the real world. Its formal definition is given as follows.

Definition 5.1 (Heterogeneous Graph) A heterogeneous graph $G_{he} = (V, E,$
$L_v, L_e, A_v, A_e)$ is a directed, labeled, and attributed graph. L_v is the label set of
vertices V and L_e is the label set of edges E. There are vertex-mapping function
$f : V \rightarrow L_v$ and edge-mapping function $g : E \rightarrow L_e$, by which each vertex and
edge are mapped to a single label individually, i.e., $f(v) \in L_v$ and $g(e) \in L_e$, where
$v \in V$ and $e \in E$. A_v and A_e are the attribute sets of vertices and edges and can be
empty.

Homogeneous graph is a special case of heterogeneous graph where $|L_v| = 1$
and $|L_e| = 1$. Here we extend the definition of homogeneous graph and introduce
edge homogeneous graph which only has $|L_e| = 1$. Since a relation only connects
two vertices, so $|L_v|$ is at most two in an edge homogeneous graph. When context
is clear, instead of edge homogeneous graph, we still use homogeneous graph for
simplicity.

5.2.2   Definition of Homogeneous Graph Extraction

Graph extraction is a process of generating graph data which is used for in-depth analysis. In this work, we focus on extracting edge homogeneous graphs from a heterogeneous graph. The edges of homogeneous graph are created based on a user input pattern on the heterogeneous graph and the attribute values of new edges are computed by user-defined aggregate functions.

Following the intuition that two objects are correlated if there exists a sequence of relations between them, the user input pattern, which defines the new relation, is a line pattern and its formal definition is given as below.

Definition 5.2 (Line Pattern) A line pattern $G_p = (V, E, L_v, L_e)$ is a directed labeled connected graph. The graph has exactly two vertices of degree one, which are named start vertex and end vertex, respectively, and the other vertices have degree two. L_v and L_e are vertex label set and edge label set, respectively. The vertex-mapping function and edge-mapping function are the same as the functions in Definition 5.1.

The subgraph instance which satisfies the line pattern is called a *path* in the heterogeneous graph. Because the path is purely determined by the line pattern, it is possible that a vertex in the path has two incoming or outgoing neighbors. Without explicit statement, the "path" is the one matched by the line pattern.

Next, on the basis of line pattern, we give the definition of homogeneous graph extraction problem.

Definition 5.3 (Homogeneous Graph Extraction Problem) Given a $G_{he} = (V, E, L_v, L_e, A_v, A_e)$, a $G_p = (V_p, E_p, L_{pv}, L_{pe})$ and user-defined aggregate functions $\otimes$ and $\oplus$ (refer to Definition 5.4), generate an edge homogeneous graph $G = (V', E', A'_e)$. Vertex set V' is the union of vertices who match the start vertex v_s and end vertex v_e of G_p, i.e., $V' = \{v | f(v) = f_p(v_s)\} \cup \{v | f(v) = f_p(v_e)\}$. Each edge $e = (u, v) \in E'$ indicates there exist at least one path matched by G_p between vertices u and v. The attribute $a \in A'_e$ of an edge $e = (u, v)$ is computed from all the paths between vertices u and v by the user-defined aggregate functions $\otimes$ and $\oplus$.

5.2.3   The Characteristics of Graph Extraction

In this part, we first introduce the two steps for solving homogeneous graph extraction problem. One is path enumeration and the other is pair-wise aggregation. Then we present the hardness of graph extraction problem.

5.2.3.1   Path Enumeration and Pair-Wise Aggregation

Path enumeration is to find out all the paths matched by the line pattern. In graph theory, a path containing no repeated vertices is a simple path, otherwise it is a non-simple path. We will prove that the extraction problem becomes intractable if we target on simple paths in Sect. 5.2.3.2.

Pair-wise aggregation computes the attribute values of edges in the extracted homogeneous graph. Each value is aggregated from two levels. First, the value of a path is aggregated from the path's edges, denoted by $\otimes$; second, the final attribute values are aggregated from the paths' values, denoted by $\oplus$. Both $\otimes$ and $\oplus$ are binary operators. The operands of $\otimes$ and $\oplus$ are the values of edges and paths, respectively. The formal definition of the two-level aggregate model is presented as below.

Definition 5.4 (Two-Level Aggregate Model) Given a heterogeneous graph G_{he}, a line pattern G_p, and user-defined aggregate functions $\otimes$ and $\oplus$, for each vertex pair u and v, P_{uv} denotes the set of paths which satisfy the line pattern G_p between the two vertices u and v, the final attribute value of edge (u, v) in the extracted homogeneous graph is computed by the following two steps. s

(1) *the value $val(p)$ of a path $p \in P_{uv}$ is*

$$val(p) = \underset{\forall e_i \in p}{\otimes} w(e_i), \tag{5.1}$$

where $w(e_i)$ is the attribute value of an edge e_i of path p.
(2) *the final attribute value $val(u, v)$ of the new edge (u, v) is*

$$val(u, v) = \underset{\forall p_i \in P_{uv}}{\oplus} val(p_i). \tag{5.2}$$

5.2.3.2   The Hardness of Graph Extraction

Next, we show the hardness of homogeneous graph extraction problem. Theorem 5.1 shows that the problem becomes intractable if path enumeration requires simple paths.

Theorem 5.1 (The Hardness of Graph Extraction with Enumerating Simple Path) *Assume the homogeneous graph extraction problem enumerates simple paths, it is #W[1]-complete[3].*[1]

Proof Counting simple paths of length k, parameterized by k, on a homogeneous graph G is #W[1]-complete, and it can be reduced to the homogeneous graph extraction problem. Given the length k and a graph G, where each edge has weight 1,

---

[1]#W[1]-complete problem does not have fixed-parameter tractable solutions.

we can construct a line pattern of length k with $|L_{pv}| = 1$ and $|L_{pe}| = 1$. The graph G is a special instance of a heterogeneous graph where $|L_v| = 1$ and $|L_e| = 1$. Then we set the user-defined aggregate functions (i.e., $\otimes$ is set to multiplication and $\oplus$ is set to addition.) for the homogeneous graph extraction problem to count the number of paths between each pair of vertices. Finally, the global number of paths can be computed from the pair-wise path count in polynomial time. So far the counting simple path problem has been successfully reduced to the homogeneous graph extraction problem. □

Considering that, in real applications with the help of diverse types of vertices and edges, there are many meaningful non-simple paths. For example, when we extract the relations of two authors publishing papers at the same venue, it is reasonable to consider authors of the same paper have such relation. Therefore, we enumerate non-simple paths, and the problem is no longer #W[1]-complete.

However, because the number of paths is exponential to the length of line pattern, the problem is still computation intensive. But we find that, both the path enumeration and pair-wise aggregation are computed based on paths' end vertices and all the intermediate paths are independent. This observation indicates that all the computation fit in vertex-centric based parallel graph processing systems. Therefore, to improve the performance, we unify the path enumeration and pair-wise aggregation into a parallel graph extraction framework, which is introduced in the following section.

5.3 Parallel Graph Extraction Framework

As discussed previously, the path enumeration and pair-wise aggregation are well suited to the vertex-centric model. Consequently, we propose a parallel graph extraction framework by using popular graph computation systems [4, 7, 11, 17]. The basic assumptions for the framework design are as follows. The heterogeneous graph is partitioned and is stored in distributed memory. The intermediate states are transferred among computing nodes by message passing.

Figure 5.3 depicts the overview of execution flow in the framework. The left part of the figure shows the logic in the master node. (1) After users submit a line pattern and aggregate functions $\otimes$ and $\oplus$, the master compiles the input line pattern into a path concatenation plan (PCP), which defines the order of concatenating short paths to generate the final paths, and sends the plan to the computing nodes. This strategy follows the idea that a long path can be generated by concatenating short paths to reduce the number of iterations. (2) On computing nodes, the framework generates complete paths by iteratively evaluating the plan in a vertex-centric manner. (3) Finally, the framework creates the extracted homogeneous graph by computing the aggregate functions $\otimes$ and $\oplus$ following the two-level aggregate model. Since the final matched paths are stored at one of its end vertices as neighbors after the path enumeration, the aggregation can be done in a vertex-centric manner as well.

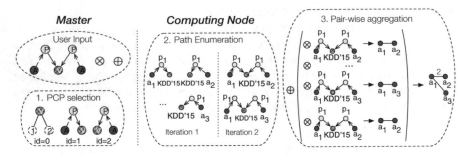

Fig. 5.3 Execution flow of parallel graph extraction framework. In the example, the master node selects a PCP with depth 2, and the computing nodes generate complete paths in two iterations and create the extracted homogeneous graph by computing the aggregate functions under the two-level aggregate model

The performance of the framework is determined by path enumeration and pair-wise aggregation. From Fig. 5.3, we notice that the number of intermediate paths and the number of iterations affect the performance. In the following subsections, we first present concepts of primitive pattern and PCP, then describe the basic version of PCP evaluation algorithm followed by the cost analysis of the framework. In Sect. 5.4, we will introduce partial aggregation to improve the performance of pair-wise aggregation by reducing the number of intermediate paths. In Sect. 5.5, we discuss techniques to select a good PCP which reduces both the number of intermediate paths and the number of iterations.

5.3.1   Primitive Pattern and Path Concatenation Plan

The goal of path concatenation plan is to decide the order of short paths which are concatenated to produce long paths. In a plan, the minimal concatenation operation is that each vertex v generates a new path by connecting one path which ends with the vertex v to another path which starts with the vertex v. This logic of concatenation operation is similar to evaluate a line pattern of length two, where the middle vertex concatenates the results matched left side and the results matched right side. Thereby, the minimal concatenation operation can be expressed as a line pattern of length two, which is called *primitive pattern*. A path concatenation plan consists of a set of related primitive patterns.

Before giving the formal definition of primitive pattern, we explain the concept of *extended vertex label set* and *extended edge label set*. Given a heterogeneous graph $G_{he} = (V, E, L_v, L_e, A_v, A_e)$, the extended vertex label set denoted by $L_{ev} = L_v \cup \{id_i\}$ includes additional vertex labels $\{id_i\}$ which are the identifications of primitive patterns; and the extended edge label set denoted by $L_{ee} = L_e \cup \{\phi\}$ includes an empty edge label ϕ which means no edge matching is required between vertices. To distinguish the new labels ($\{id_i\}$ and ϕ) from the original labels (L_v and L_e),

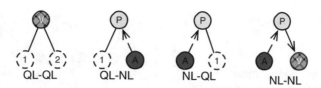

Fig. 5.4 Four types of primitive patterns on the scholarly graph. The leftmost figure shows that v_s matches the results of primitive pattern with $id = 1$; v_e matches the results of primitive pattern with $id = 2$; both matched results are concatenated by Venue vertex v_p. The other three figures have the similar explanations

we name the new labels as **query labels** denoted by QL, and the original labels are **native labels** denoted by NL. The following definition describes the concept of primitive pattern.

Definition 5.5 (Primitive Pattern) The primitive pattern denoted by $P_p = (id, V_{pp}, E_{pp}, L_{ev}, L_{ee})$ is a line pattern of length two with extended vertex and edge label sets, and has a unique identification id. The vertex set V_{pp} consists of three vertices, start vertex v_s, pivot vertex v_p, and end vertex v_e, i.e., $V_{pp} = \{v_s, v_p, v_e\}$. The edge set E_{pp} includes a left edge $e_l = (v_s, v_p, dir)$ and a right edge $e_r = (v_p, v_r, dir)$, where dir is the direction (incoming, outgoing, or undirected) of the edge, i.e., $E_{pp} = \{e_l, e_r\}$. The mapping functions are similar to the functions in Definition 5.2.

With respect to the different combinations of vertex labels, there are four types of valid primitive patterns. They are NL-NL pattern, NL-QL pattern, QL-NL pattern, and QL-QL pattern. An NL-QL pattern means $f(v_s)$ is a native label (i.e., L_v, L_e) and $f(v_e)$ is a query label (i.e., $\{id_i\}$, ϕ). Figure 5.4 shows four examples of different types of primitive patterns on the scholarly graph.

Now a general line pattern can be compiled into a path concatenation plan expressed by a set of primitive patterns. The formal definition of the plan is stated as below.

Definition 5.6 (Path Concatenation Plan (PCP)) PCP is a collection of primitive patterns. The dependence between the primitive patterns forms a binary tree, where each node is a primitive pattern and the leaf nodes are NL-NL patterns. In the tree, a node X is the parent of a node Y only if Y's primitive pattern id is the vertex label in X's primitive pattern.

The height of a PCP tree is denoted by H, and from root to leaf, the level number of each node is one to H. The node close to the root has small level number. Figure 5.5 lists five different PCPs of a single line pattern. PCP-1 consists of three primitive patterns and is a binary tree of height 2. The primitive pattern $id = 0$ is the root node with level number 1, and the other two primitive patterns are leaf nodes which are both NL-NL patterns. Except PCP-1, the other four PCPs all have height of three. The following theorem shows the lower bound of the height of a PCP for a line pattern.

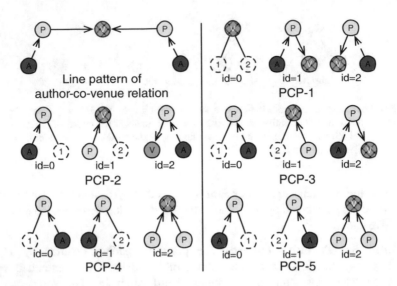

Fig. 5.5 Five different PCPs of a single line pattern which defines the relation of two authors publishing papers at the same venue

Theorem 5.2 *Given a line pattern of length l (l $\geq$ 2), the height of a PCP is at least $\lceil log_2(l) \rceil$.*

Proof A line pattern of length l generates a PCP with $l - 1$ nodes. Because a balanced full binary tree with n nodes has height $\lceil log_2(n + 1) \rceil$, the height of a PCP of a line pattern of length l is at least $\lceil log_2(l) \rceil$.   □

5.3.2   PCP Evaluation Algorithm

Intuitively, a PCP should be evaluated from leaf to root. Furthermore, primitive patterns in the same level can be evaluated simultaneously, since they are independent of each other. The core idea of evaluating a PCP is iteratively computing an array of primitive patterns in the same level, and the complete paths will not be generated until the computation reaches the root primitive pattern.

The pseudocode of the evaluation algorithm is illustrated in Algorithm 1. Given the fact that the length of a primitive pattern is two, both the start vertex and end vertex are the neighbors of pivot vertex. This property allows us to concatenate partial paths on pivot vertex by neighbor exploring instead of global searching. Therefore, to evaluate a primitive pattern, we use the vertex-centric model [7] which is easily parallelized. More specifically, in an iteration (Lines 4–11), each vertex in graph G_{he} generates paths for primitive patterns and sends the paths to the corresponding vertices by calling Algorithm 2. During the concatenation, the partial

---

Algorithm 1 PCP evaluation pseudocode

Input: PCP *plan*, The height of PCP H, Aggregate functions $\otimes$ and $\oplus$, Heterogeneous graph G_{he}
Output: The extracted homogeneous graph G_{ho}
  1: **foreach** vertex v in G_{he} **do**
  2:     preprocessing of materializing both out/in neighbors of v at local;
  3: **end foreach**{preprocessing}
  4: **foreach** $h \leftarrow H$ to 1 **do**
  5:     $ppSet \leftarrow plan.getPPbyLevel(h)$;
  6:     **foreach** vertex v in G_{he} **do**
  7:         **foreach** primitive pattern pp in $ppSet$ **do**
  8:             evaluate pp on vertex v by Algorithms 2;
  9:         **end foreach**
 10:     **end foreach**
 11: **end foreach**{path enumeration}
 12: **foreach** vertex v_e in G_{he} **do**
 13:     $PS \leftarrow$ get paths end with v_e and index the paths by their start vertices v_s;
 14:     **foreach** vertex v_s **do**
 15:         $path_{v_s \rightarrow v_e} \leftarrow$ get paths start with v_s from PS;
 16:         $aggVal \leftarrow \phi$;
 17:         **foreach** path pa in $path_{v_s \rightarrow v_e}$ **do**
 18:             $pathVal \leftarrow \otimes w(e_i), e_i \in pa$;
 19:             $aggVal \leftarrow val \oplus pathVal$
 20:         **end foreach**
 21:         insert extracted relation $(v_s, v_e, aggVal)$ into G_{ho}.
 22:     **end foreach**
 23: **end foreach**{pair-wise aggregation}
 24: **return** G_{ho};

---

paths for matching start vertex and end vertex of a primitive pattern come from two different sources in accordance with the labels of vertices. The vertex with NL label directly matches the data of the heterogeneous graph, while the vertex with QL label matches the results of previous primitive pattern. For example, in Fig. 5.3, iteration 1 processes NL-NL primitive patterns $id = 1$ and $id = 2$, which access the data of G_{he} (Lines 5 and 10 in Algorithm 2), and the iteration 2 processes QL-QL primitive pattern $id = 0$, which uses the partial paths generated by primitive patterns $id = 1$ and $id = 2$ (Lines 7 and 12 in Algorithm 2).

However, when matching the vertex with NL label, because the edges in a primitive pattern have directions, the pivot vertex needs to explore both in and out neighbors. Considering graph computation systems only maintains out neighbors for each vertex, we bring in a preprocessing phase (Lines 1–3 in Algorithm 1), in which each vertex materializes its in and out neighbors at local, to truly achieve the neighbor exploration of pivot vertex. Finally, benefit from the vertex-centric model, all the complete paths generated by the PCP evaluation are stored at their end vertices, respectively. Therefore, the pair-wise aggregation (Lines 12–23 in Algorithm 1) can be naturally executed in the vertex-centric model as well.

Algorithm 2 Primitive pattern evaluation pseudocode

Input: Vertex $v \in G_{he}$, Primitive Pattern pp
 1: initialize v_s, v_p, v_e, e_l, e_r by pp;
 2: **if** $f(v) = f_p(v_p)$ **then**
 3:     $leftRes \leftarrow NULL$; $rightRes \leftarrow NULL$;
 4:     **if** $f_p(v_s) \in NL$ **then**
 5:         $leftRes \leftarrow$ matching e_l to the v's neighbors in G_{he};
 6:     **else**
 7:         $leftRes \leftarrow$ retrieving the results of primitive pattern $id = f_p(v_s)$;
 8:     **end if**
 9:     **if** $f_p(v_e) \in NL$ **then**
10:         $rightRes \leftarrow$ matching e_r to the v's neighbors in G_{he};
11:     **else**
12:         $rightRes \leftarrow$ retrieving the results of primitive pattern $id = f_p(v_e)$;
13:     **end if**
14:     concatenate $leftRes$ and $rightRes$;
15:     **if** pp is the left child **then**
16:         send the new paths to the end vertex;
17:     **else**
18:         send the new paths to the start vertex;
19:     **end if**
20: **end if**

5.3.3   Cost Analysis

According to Algorithms 1 and 2, the total computation cost is mainly related to the following three parts. First, the cost of accessing each vertex in the heterogeneous graph (Line 6 in Algorithm 1); second, the cost of concatenating paths (Algorithm 2); third, the cost of pair-wise aggregation. Since the preprocessing can be done off-line once, we do not count its cost into the total computation cost.

In each iteration, the algorithm scans over all the vertices in the heterogeneous graph once. Therefore, **the number of iterations** is proportion to the cost of enumerating vertices, i.e., the cost is cVH, where c is a coefficient and V is the number of vertices.

The path enumeration cost denoted by S_{pcp} is simply the sum of the cost of primitive pattern evaluation, i.e.,

$$S_{pcp} = \sum_{pp_i} S_{pp_i}, \tag{5.3}$$

where S_{pp_i} is the cost of evaluating a primitive pattern pp_i. In Algorithm 2, assume the sizes of $leftRes$ and $rightRes$ are d_{vl} and d_{vr} individually with respect to the vertex v, then S_{pp_i} is

$$S_{pp_i} = \sum_{v:f(v)=f_p(v_p)} d_{vl} * d_{vr}. \tag{5.4}$$

From the above equations, we see that S_{pp_i} is highly related to **the number of intermediate paths**. Moreover, the cost of pair-wise aggregation is proportional to the number of final paths as well.

In summary, the overall performance of parallel homogeneous graph extraction framework is highly related to two factors, the number of iterations and the number of intermediate paths. Therefore, we design optimization techniques by carefully considering the two factors.

5.4 Aggregation in Homogeneous Graph Extraction

The performance of pair-wise aggregation is proportional to the number of paths. To reduce the number of paths, we borrow the idea of classifying the aggregate functions in traditional OLAP [5], i.e., the pair-wise aggregations is divided into three types, including distributive aggregations, algebraic aggregations, and holistic aggregations. We first present the criteria to distinguish three types of aggregations; then we introduce optimization techniques for the distributive and algebraic aggregations.

5.4.1 Distributive, Algebraic, and Holistic Aggregation

Most of aggregation strategies can be represented by the two-level aggregate model (in Sect. 5.2.3). As illustrated in Algorithm 1, a universal solution for the pair-wise aggregation first exhaustively enumerates all the paths, and then applies the aggregate model on the paths. However, the number of paths is exponential to the length of line pattern, and hence it is inefficient to aggregate the values by naive enumeration.

We classify the aggregations into three different types by analyzing the properties of aggregate functions $\otimes$ and $\oplus$. They are distributive aggregations, algebraic aggregations, and holistic aggregations. The distributive aggregation can be solved in a divide-and-conquer fashion which means the final aggregation can be computed from several sub aggregations that are calculated based on partial paths. The algebraic aggregation can be solved by maintaining several distributive aggregations. A holistic aggregation needs to enumerate all the paths. In practice, the distributive aggregation is more attractable and is possible to be solved efficiently. The following theorem gives the condition to be a distributive aggregation.

Theorem 5.3 (Distributive Aggregation Condition) *If $\otimes$ is distributive over $\oplus$, then the two-level aggregation is the distributive aggregation. In other words, the*

Algorithm 3 Primitive pattern evaluation with partial aggregation

---

Input: Vertex $v \in G_{he}$, Primitive Pattern pp

 1: Lines 1–13 in Algorithm 2.
 2: /*Because of partial aggregation, each element in $leftRes$ and $rightRes$ only has three fields, matched start vertex v'_s, matched end vertex v'_e and partial aggregate value val.*/
 3: $leftAggMap \leftarrow \phi, rightAggMap \leftarrow \phi$.
 4: **foreach** $p_l = (v'_{sl}, val_l, v'_{el})$ in $leftRes$ **do**
 5:     $leftAggMap[v'_{sl}] \leftarrow leftAggMap[v'_{sl}] \oplus val_l$
 6: **end foreach**
 7: **foreach** $p_l = (v'_{sr}, val_r, v'_{er})$ in $rightRes$ **do**
 8:     $rightAggMap[v'_{er}] \leftarrow rightAggMap[v'_{er}] \oplus val_r$
 9: **end foreach**
10: **foreach** v_{sl} in $leftAggMap$ **do**
11:     **foreach** v'_{er} in $rightAggMap$ **do**
12:         $nVal \leftarrow leftAggMap[v'_{sl}] \otimes rightAggMap[v'_{er}]$.
13:         new partial path $nPath = (v'_{sl}, nVal, v'_{er})$.
14:         Lines 15–19 in Algorithm 2.
15:     **end foreach**
16: **end foreach**

---

equation below holds.

$$val_l(u, v) = \bigoplus_{\forall v_t \in V} (val_t(u, v_t) \otimes val_{l-t}(v_t, v)), \tag{5.5}$$

where $val_l(u, v)$ is the aggregate value from paths of length l between vertices u and v, $t < l$ and v_t is the tth vertex in a path of length l.

Proof Equation 5.5 can be deducted as follows by the property that $\otimes$ is distributive over $\oplus$.

$$val_l(u, v) = \bigoplus_{p_i \in P_{uv}} (\bigotimes_{e_{ij} \in p_i} w(e_{ij})) \quad \because \text{Definition 5.4.}$$

$$= \bigoplus_{\forall v_t \in V} (\bigoplus_{p_i(t)=v_t \wedge p_i \in P_{uv}} \bigotimes_{e_{ij} \in p_i} w(e_{ij}))$$

$$\because \text{group } p_{uv} \text{ by the } t\text{th vertex } v_t.$$

$$= \bigoplus_{\forall v_t \in V} ([\bigoplus_{p_i \in P_{uv_t}(t)} \bigotimes_{e_{ij} \in p_i} w(e_{ij})]$$

$$\otimes [\bigoplus_{p_i \in P_{v_t v}(l-t)} \bigotimes_{e_{ij} \in p_i} w(e_{ij})])$$

$$\because \otimes \text{ is distributive over } \oplus.$$

$$= \bigoplus_{\forall v_t \in V} (w_t(u, v_t) \otimes w_{l-t}(v_t, v)). \tag{5.6}$$

$\square$

According to Theorem 5.3, a lot of common aggregate functions are distributive aggregations. Name a few, multiplication-addition ($\otimes$-$\oplus$), min-max, max-min, addition-max, sum-min, etc. The computation of these aggregate functions can be optimized by aggregating during the path enumeration.

Since algebraic aggregation is a combination of distributive aggregation, it can be benefited from Theorem 5.3 as well. The holistic aggregations are still computation and space expensive. In this work, we focus on optimizing the computation of distributive and algebraic aggregations, because the majority of analysis scenarios [2, 6, 19, 20, 22] handle these two kinds of aggregations. Furthermore, the optimization of holistic aggregation requires special designs according to its concrete aggregate functions [8].

5.4.2   Optimization with Partial Aggregation

Theorem 5.3 points out that distributive aggregations can be aggregated ahead with partial paths. In other words, distributive aggregations can be computed through partial aggregation. The partial aggregation merges intermediate paths with the same start and end vertices by the aggregate functions before the complete paths are found. As a result, the number of intermediate paths is reduced and the overall performance is improved.

The algorithm with partial aggregation is illustrated in Algorithm 3. Before generating new paths, we first use $\oplus$ to merge the paths with the same start and end vertices from different workers (Lines 3–9 in Algorithm 3). In Lines 10–15, when creating the new paths, we also use $\otimes$ to compute the partial aggregated values for the new partial paths. With above optimizations, the number of intermediate paths in each iteration is no longer exponential, and it is bounded by $O(V^3)$, where V is the number of vertices in the heterogeneous graph. In addition, with the partial aggregation, for each path, we only need to store and transfer start vertex, end vertex, and the (partially) aggregate value, so the memory and communication costs are reduced compared to Algorithm 2 as well.

5.5   Path Concatenation Plan Selection

As introduced in Sect. 5.3.1, a line pattern can be decomposed into many different PCPs (refer to Fig. 5.5). Each PCP has different heights and generates different sizes of intermediate paths during the computation which results in different performance. According to the cost analysis in Sect. 5.3.3, a better PCP should incur less number of intermediate paths and less number of iterations. However, the number of iterations and the size of intermediate paths do not have positive correlation. Therefore, in the following sections, we mainly discuss three strategies for selecting a good PCP. They are iteration optimized strategy, path optimized strategy, and hybrid strategy.

5.5.1 The Path Size Estimation for PCP

Before introducing the details of PCP selection, we present an approach to estimate the path size for a given PCP. From Eq. 5.4, terms $\sum_{v:f(v)=f(v_p)} d_{vl}$ and $\sum_{v:f(v)=f(v_p)} d_{vr}$ are the size of intermediate paths generated by the children of pp_i, when the pp_i has QL labels; otherwise, the terms are the number of edges in the heterogeneous graph. Hence, to compute S_{pp_i}, we need to know the distribution of intermediate paths and edges with regard to the vertices of heterogeneous graph.

Given a primitive pattern pp_i, assume that pp_j is the left child of pp_i when v_s has QL label, otherwise L denotes the number of edges matched by pp_i's left sides; pp_k and R denote the right child of pp_i and the number of edges matched by pp_i's right side, respectively. Then under the uniform distribution model (i.e., the edges and intermediate paths are uniformly distributed over the vertices of heterogeneous graph), we can extend Eq. 5.4 into the following model to estimate the cost of a pp_i in a PCP,

$$
S_{pp_i} = \begin{cases}
\frac{S_{pp_j} \times S_{pp_k}}{|V_p|} & QL - QL \text{ pattern} \\
\frac{L \times S_{pp_k}}{|V_p|} & NL - QL \text{ pattern} \\
\frac{S_{pp_j} \times R}{|V_p|} & QL - NL \text{ pattern} \\
\frac{L \times R}{|V_p|} & NL - NL \text{ pattern}
\end{cases}
\tag{5.7}
$$

where V_p is the vertex set in which each vertex matches the pivot vertex of pp_i. According to Eq. 5.3, the estimated cost of a PCP (S_{pcp}) can be computed by summing up Eq. 5.7 of different pp_is.

Moreover, a sophisticated distribution assumption (e.g., power law or normal distributions) can be used to increase the accuracy of the estimation. However, similar to the work [12], it requires other techniques to estimate the parameters of the distribution, which is not the focus of our work. Through the experiments, we can observe that the uniform distribution assumption is fair enough to help us select a good plan.

5.5.2 PCP Selection

In this subsection, we elaborate three strategies of PCP selection. The first strategy is iteration optimized strategy which selects a PCP with the minimized number of iterations. The second strategy is path optimized strategy which selects a PCP with minimized size of intermediate paths. The third one is a hybrid strategy which selects a PCP with minimized size of intermediate paths from PCPs with minimized number of iterations.

5.5.2.1 Iteration Optimized Strategy

Theorem 5.2 shows that the PCPs of a line pattern of length l have heights at least $\lceil log(l) \rceil$. This lower bound is achieved when PCP is a balanced full binary tree. Thereby, we define the iteration optimized strategy as below.

Definition (Iteration Optimized Strategy) Given a line pattern P of length l, the goal of this strategy is to select out a PCP which has $\lceil log(l) \rceil$ height.

The main procedure for generating such a PCP is that we use divide-and-merge framework to recursively divide the line pattern into two sub line patterns with the same length until each sub line pattern has length less than three; then merge these sub line patterns to form a PCP. During the division phase, there might be multiple vertices which divide the line pattern into two sub line patterns with the same length. In such cases, we randomly pick one. The procedure is finished in $O(l)$ time. This strategy only reduces the number of iterations without considering the size of intermediate paths.

5.5.2.2 Path Optimized Strategy

The goal of path optimized strategy is to find out the best PCP which generates the minimized number of intermediate results. The optimization problem is presented as below.

Definition 5.7 (Path Optimized Strategy) Given a line pattern P and a heterogeneous graph G_{he}, the goal of the problem is to find a PCP satisfying that

$$\min\{S_{pcp}\}.$$

Each PCP forms a binary tree (Definition 5.6), different PCPs may share the same subtrees. For example, in PCP-3 and PCP-4 shown by Fig. 5.5, the left vertices in their primitive pattern $id = 0$ are corresponding to the results from the same sub pattern, which consists of the first four vertices of the original line pattern. This observation leads to the optimal substructure for the path optimized strategy, which is formalized by the following equations. Given a line pattern of length l, the vertices are sequentially numbered from left to right, then

$$S_{pcp}[i, j] = \begin{cases} \min\limits_{i<k<j} \{S_{pcp}[i, k] + S_{pcp}[k, j] + S_{pp_k}\} & j - i > 2 \\ 0 & j - i \leq 2 \\ , \end{cases}$$

where i, j, k are the ids of vertices in the line pattern and $S_{pcp}[i, j]$ is the cost of a sub pattern of line pattern between vertices i and j. In addition, k implies that the vertex k to be a pivot vertex in pp_k. Consequentially, the best PCP has the cost $S_{pcp}[1, l]$. The above equations can be solved in $O(l^3)$ by the dynamic programming technique. This strategy only optimizes the size of intermediate results without considering the number of iterations.

5.5.2.3   Hybrid Strategy

As mentioned at the beginning, the number of iterations and the size of intermediate paths do not have positive correlation. In practice, the number of iterations has more influence than the size of intermediate paths on the overall performance. This is because the distribution of intermediate paths is hard to be predicated, and they cannot be processed by the workers in balance. Therefore, increasing the number of intermediate paths to some extend may improve the utility of workers, and the overall performance will not degrade significantly. In contrast, by applying the hash partition schema, the vertices of heterogeneous graph can be evenly distributed over the workers, increasing the number of iterations certainly degrades the overall performance.

Based on above observations, the PCP selection strategy should give priority to reduce the number of iterations. More specifically, a PCP should have the number of iterations as small as possible before reducing the size of intermediate paths. So we design a hybrid strategy by using both iteration optimized strategy and path optimized strategy. We use the iteration optimized strategy to construct a balanced binary tree, but compared to the previous random selection of pivot vertices, we choose the pivot vertices according to the cost model with dynamic programming technique. The new dynamic programming equations are presented as below.

$$S_{pcp}[i, j] = \begin{cases} \min\limits_{k \in \{\lfloor \frac{i+j}{2} \rfloor, \lceil \frac{i+j}{2} \rceil\}} \{S_{pcp}[i, k] + S_{pcp}[k, j] + S_{pp_k}\} & \\ & j - i > 2 \\ 0 & j - i \leq 2 \end{cases}.$$

Compared to the path optimized strategy, the dynamic programming equations only enumerate pivot vertices (k) around the center index in the line pattern to guarantee the final PCP minimizes number of iterations.

5.6 Experiments

In this section, we first demonstrate the effectiveness of plan selection algorithm and show the benefit of partial aggregation techniques. Then we compare our solution with three different approaches, they are based on graph database, regular path query (RPQ), and matrix model, respectively. Finally we show the scalability of our solution with respect to the number of workers, the size of datasets, and the length of line pattern.

5.6.1 Experiment Settings

The prototype of parallel graph extraction framework was implemented on top of Giraph.[2] The framework reads a line pattern from HDFS and the line pattern is represented by a line graph. The user-defined aggregate functions are two new abstract methods (i.e., $\otimes$ and $\oplus$) in the vertex program of primitive pattern evaluation, which uses Algorithm 2 for the basic extraction solution, and Algorithm 3 for the extraction solution with partial aggregation. Besides, the framework involves a preprocessing phase as well. The preprocessing aims to materialize in and out neighbors of each vertex with the corresponding labels. This can be finished in three iterations and only conducted once.

The experiments were conducted on a cluster with 22 physical nodes, who have 48G memory and a 2.6Ghz CPU individually. We use two datasets, dblp-2014[3] and US-patent,[4] for the experiments. The schemes of dblp-2014 and US-patent are shown in Figs. 5.6a and 5.7a.

Line Patterns The design of line pattern requires domain knowledge, because different analysis applications have different demands on the extracted graph and aggregate functions. However, according to many existing applications and literature [6, 19], the line pattern can be still classified into two categories.

- Bipartite line pattern (BP for short): the pattern defines new relations of two different types of vertices.
- Symmetry line pattern (SP for short): the pattern defines new relations between the same type of vertices.

On dblp-2014 dataset, we define three symmetry patterns and one bipartite pattern. They are explained as follows.

---

[2]http://giraph.apache.org/.

[3]http://dblp.uni-trier.de/xml/.

[4]http://www.nber.org/patents/.

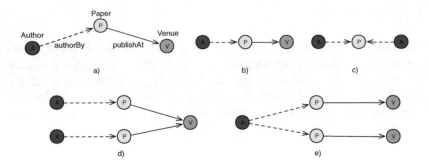

Fig. 5.6 dblp-2014 schema and line patterns. (**a**) dblp-2014 schema. (**b**) authorVenue BP. (**c**) coAuthor SP. (**d**) coAuthorVenue SP. (**e**) coVenueAuthor SP

- dblp-BP1 (Fig. 5.6b): the pattern extracts publish relation between authors and venues.
- dblp-SP1 (Fig. 5.6c): the pattern extracts co-authorship among authors.
- dblp-SP2 (Fig. 5.6d): the pattern extracts the relation of authors who publish papers on the same venue.
- dblp-SP3 (Fig. 5.6e): the pattern extracts the relation of venues where papers of the same author are published.

On US-patent dataset, we define three symmetry patterns and two bipartite patterns on US-patent. The details are listed as below.

- patent-BP1 (Fig. 5.7e): a bipartite pattern extracts relation between locations and categories of the patents.
- patent-BP2 (Fig. 5.7f): a bipartite pattern extracts the two-hop citation relation between the inventor and patents.
- patent-SP1 (Fig. 5.7b): the pattern extracts co-inventor relation among inventors.
- patent-SP2 (Fig. 5.7c): a symmetry pattern extracts the citation relation among different locations.
- patent-SP3 (Fig. 5.7d): a symmetry pattern extracts the citation relation among inventors.

According to the size of results of each pattern, we also divide above patterns into light patterns and heavy patterns which are listed in Table 5.1.

Aggregate Functions Since different instances of distributive and algebraic aggregations have the similar computation patterns, in the experiments, we use path counting as a representative.

Baselines We compared our prototype with three different kinds of approaches. First one is graph database-based approach. We use graph database Neo4j[5] to

---

[5]http://neo4j.com/.

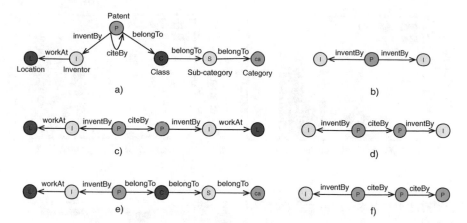

Fig. 5.7 US-patent schema and line patterns. (**a**) US-patent schema. (**b**) coInventor SP. (**c**) coLocationCitation SP. (**d**) coInventorCitation SP. (**e**) locationCategory BP. (**f**) inventorPatentCitation BP

Table 5.1 Light and heavy patterns

Type	Line pattern	No. of inter. paths
Light	patent-BP1	5,286,917
	dblp-BP1	7,480,864
	patent-SP1	12,972,984
	dblp-SP1	28,384,288
	patent-SP3	50,397,805
Heavy	patent-SP2	81,252,106
	dblp-SP3	359,793,482
	patent-BP2	545,208,404
	dblp-SP2	9,760,368,782

store the heterogeneous graphs, and ran the database on a Linux server with 96GB memory. Since graph database provides the interface of querying paths from a certain vertex. By using the graph database to answer a query, we first retrieve vertices matched by the start vertex of the input pattern; then we query the paths and aggregate them for each retrieved vertex.

Second one is matrix-based solution which is introduced by Rodriguez [13]. The path enumeration and pair-wise aggregation processing can be converted into matrix multiplications. A heterogeneous network is transformed into a set of matrixes and vertex mappings. To answer a query, we first retrieve related matrixes and do the matrix multiplications, then transform the results into a subgraph in the original heterogeneous network by using the vertex mappings. This solution is implemented based on the sparse matrix multiplication in SciPy.[6]

---

[6]http://docs.scipy.org.

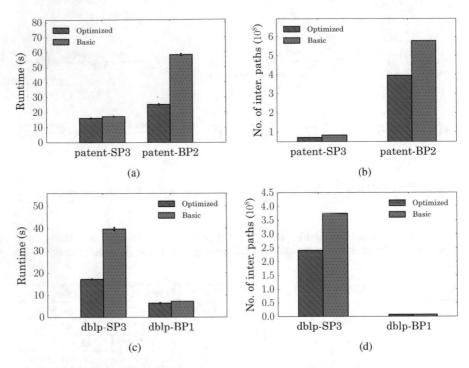

Fig. 5.8 Comparison of runtime and the number of intermediate paths between basic and optimized homogeneous graph extraction solutions. (**a, b**) US-patent. (**c, d**) dblp-2014

Third one is RPQ-based solution which is implemented following the idea in work [10].

Each experiment is ran five times, and the average performance and error-bar are reported.

5.6.2   *Effectiveness of Partial Aggregation Technique*

We first demonstrate the effectiveness of partial aggregation technique through comparing the basic and optimized graph extraction solutions. The **basic solution** first enumerates all the paths that matched by the line pattern and then computes the aggregate functions. The **optimized solution** aggregates partial paths during the path enumeration. Both solutions use hybrid strategy to select a PCP. As representatives, we show the results of running dblp-SP3 and dblp-BP1 on dblp-2014 dataset and patent-SP3 and patent-BP2 on US-patent dataset with ten workers.

Figure 5.8 shows the runtime and the number of intermediate paths for each pattern. As discussed in Sect. 5.3.3, the larger number of intermediate paths leads to poorer overall performance. Since the optimized solution reduces the number of intermediate paths with the partial aggregation technique, the performance for each pattern shall be improved. For example, running dblp-SP3, the basic solution takes about 40 s with processing about 374 million paths, while the optimized one only takes about 17 s and the number of intermediate paths is around 241 million. In addition, when the number of intermediate paths is close, both solutions have similar performance. The optimized solution outperforms a bit because it executes aggregation $\otimes$ during the path enumeration. The dblp-BP1 is an example of such case.

5.6.3 Comparison of Different Plans

According to the cost analysis in Sect. 5.3.3, the performance of homogeneous graph extraction solution is related to the number of intermediate paths and the number of iterations. We evaluate four types of plan selection strategies, line strategy (line), iteration optimized strategy (iterOPT), path optimized strategy (pathOPT), and hybrid strategy (hybrid). The line strategy indicates that we enumerate paths by expanding the pattern from one end vertex to the other end vertex in sequence. The other three strategies have been introduced in Sect. 5.5. In addition, the pathOPT and hybrid strategies select PCPs according to the estimation model proposed in Sect. 5.5.1. The following experiments are conducted by using the partial aggregation technique with ten workers.

Figure 5.9 presents the performance of three patterns (patent-SP2, patent-BP1, and dblp-SP3) on US-patent and dblp-2014 datasets as representatives. Other patterns have similar results to the one of dblp-SP3. It is clear to see from Fig. 5.9a that the hybrid strategy has the best performance across different patterns and datasets. This is because the hybrid strategy synthetically considers the number of iterations and the size of intermediate paths when selecting a plan. For example, when running patent-SP2, the hybrid strategy and pathOPT strategy select the same PCP, and they have similar performance. The iterOPT strategy has the same number of iterations (Fig. 5.9c), but has larger number of intermediate paths (Fig. 5.9b), so it performs a bit worse compared to the hybrid and pathOPT strategies. When running patent-BP1, though line and pathOPT strategies lead to small size of intermediate paths (Fig. 5.9b), they lose high performance because of the large number of iterations (Fig. 5.9c). In addition, for other patterns used in experiments, different plan selection strategies choose the same PCP, so the performance is similar. The result of dblp-SP3 is visualized as a representative.

From all the experiments, the line strategy always has the worst performance. This indicates that, without optimizing the number of iterations and the size of intermediate paths, the solution cannot be efficient. In general, the hybrid strategy is the best option to efficiently execute the extraction framework. Moreover, in Fig. 5.9b, we noticed that pathOPT and hybrid usually generate PCPs with the size of intermediate paths smaller compared to other approaches, which confirms the effectiveness of our estimation model.

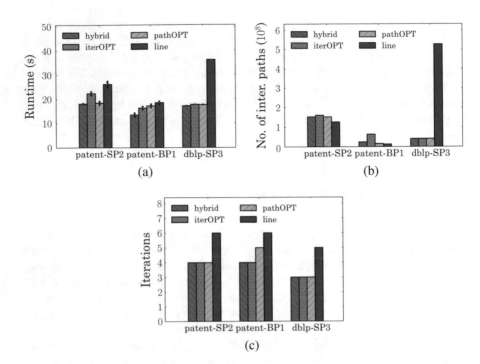

Fig. 5.9 Performance comparison of different plan selection strategies. (**a**) Runtime. (**b**) #intermediate paths. (**c**) #iterations

Table 5.2 Performance (seconds) comparison among PGE, graph database-based approach, and matrix-based approach

Pattern	Single-PGE	Graph database	Matrix-based
patent-SP1	43.74	3938.38	**19.86**
patent-SP2	88.48	1771.14	**7.01**
patent-SP3	**70.18**	5173.41	76.7
patent-BP1	60.50	149.10	**1.01**
patent-BP2	**357.07**	11,181.98	410.3
dblp-SP1	**36.49**	4116.83	48.6
dblp-SP3	**103.58**	7183.39	202.2
dblp-BP1	31.17	4249.38	**19.95**

The best performance is highlighted in bold style

5.6.4 Comparison of Standalone Solution

Here we present the results of comparing our parallel graph extraction solution (PGE for short) with two standalone solutions, graph database-based solution and matrix-based solution, in Table 5.2. We run PGE with hybrid plan in a single worker.

Comparing with the graph database-based solution, we clearly see that, even with a single worker, the PGE has a better performance. This is because, the graph database is only optimized for querying local graph data, and it cannot achieve high performance when handling global graph data.

The performances of Matrix-based solution and PGE may vary for different queries. After profiling, we found that when the final matrix is small or sparse, matrix-based solution performs better, otherwise PGE wins. First, this is because the time cost of transforming the final matrix into the subgraph with original vertex ids is proportional to the number of edges in the matrix. Second reason is that, PGE is ran in a parallel framework with a single worker, the overhead is much larger than the matrix-based solution.

5.6.5 Comparison of RPQ-Based Solution

Now we show the results of comparing our parallel graph extraction solution to the RPQ-based solution [10] in Table 5.3. For the parallel graph extraction solution, we use the hybrid plan. And both solutions are run in parallel with ten works. The results show that current RPQ solution has good performance when the extraction workload is light and it performs bad when the workload increases. For example, when running dblp-SP3, our solution is about two times faster than the RPQ-based solution. This is because the RPQ sequentially evaluate the line pattern based on the regular expression, and this type of evaluation leads to large number of iterations and does not optimized the size of intermediate results.

5.6.6 Scalability

Finally, we conduct experiments to evaluate the scalability of the parallel homogeneous graph extraction approach. The experiments demonstrate that the approach scales well with increasing the number of workers, increasing the size of dataset, or increasing the length of line pattern.

Scalability with Varying the Number of Workers Figure 5.10a shows the runtime of executing dblp-SP2 with different number of workers. The results reveal

Table 5.3 The performance (seconds) comparison of RPQ-based solution and PGE with hybrid plan

Type	Line pattern	PGE-hybrid	RPQ
Light	patent-BP1	13.60	10.19
	dblp-BP1	6.38	4.79
	patent-SP1	10.68	8.35
	dblp-SP1	7.03	5.84
	patent-SP3	15.43	15.38
Heavy	patent-SP2	17.96	30.75
	dblp-SP3	19.04	46.38
	patent-BP2	29.21	46.41
	dblp-SP2	289.08	OOM

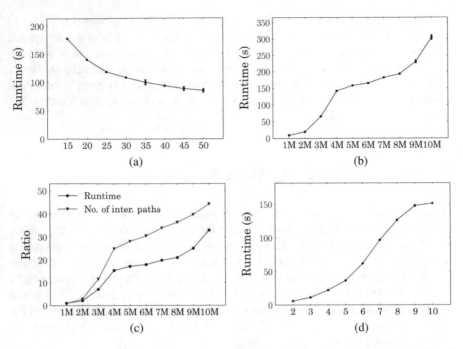

Fig. 5.10 Scalability test. (**a**) Varying the No. of workers. (**b**) Runtime with varying the No. of vertices. (**c**) Ratio with varying the No. of vertices. (**d**) Varying the length of line pattern

that the solution almost scales linearly when the number of workers increases. For example, the dblp-SP2 spends about 140 s in extracting the homogeneous graph when 20 workers are provided; and the cost is reduced to about 93 s with doubling the number of workers (i.e., 40).

Scalability with Varying the Size of Dataset Figure 5.10b and c illustrates the results of running dblp-SP2 on datasets with varying the number of vertices from 1M to 10M. Since the original dblp-2014 only contains about 4M vertices, we generate a set of synthetic datasets in different size as follows. The datasets whose number of vertices is less than 4M are generated by randomly sampling vertices from the original dblp-2014. We generate datasets having more than 4M vertices by adding new fake venues, which are randomly sampled from the existing venues. From Fig. 5.10b, we can see that the runtime increases when the size of dataset becomes large, but the increase is not linear to the size of dataset. As analyzed previously, the performance of our solution is affected by the number of intermediate paths when the number of iterations is the same. We further profiled the size of intermediate paths for different datasets. And the sizes of intermediate paths on different datasets are normalized to the size of intermediate paths on dataset with 1M vertices. So does the runtime. Figure 5.10c illustrates the normalized values (or ratios), and it shows that the degradation of the performance is proportion to the number of intermediate paths. This demonstrates that the solution still scales well to the size of dataset implicitly.

Scalability with Varying the Length of Line Pattern Increasing the length of line pattern, the number of intermediate paths increases exponentially. However, by using partial aggregation technique, the actual size of intermediate paths is polynomial. Figure 5.10d shows the results of running line patterns of different lengths with 40 workers on US-patent dataset. Here we use the citeBy relation to generate line patterns of different lengths. A line pattern of length l means that the line pattern consists of l citeBy relations. The results demonstrate that when the length of line pattern increases at the beginning, the performance degrades fast because of the fast increasing of the number of intermediate paths; but when the length of line pattern exceeds a certain threshold, like nine in the experiment, the decrease of the performance becomes slightly. This is due to the benefit brought by the partial aggregation technique.

5.7 Summary

Heterogeneous graph is a natural model to describe the complex relations in the real world. We focused on studying the problem of extracting homogeneous graph from a heterogeneous graph based on the user input pattern. To efficiently solve the problem, we proposed a parallel graph extraction algorithm with the help of path concatenation strategy. Demonstrated by the extensive experiments, the proposed solution indeed extracted homogeneous graphs efficiently compared to graph-based and matrix-based solutions.

References

1. Deng Cai, Zheng Shao, Xiaofei He, Xifeng Yan, and Jiawei Han. Community mining from multi-relational networks. In *PKDD*, pages 445–452, 2005.
2. Chen Chen, X. Yan, Feida Zhu, Jiawei Han, and P.S. Yu. Graph OLAP: Towards online analytical processing on graphs. In *ICDM*, pages 103–112, 2008.
3. Jörg Flum and Martin Grohe. The parameterized complexity of counting problems. *SIAM J. Comput.*, 33(4):892–922, April 2004.
4. Joseph E. Gonzalez, Yucheng Low, Haijie Gu, Danny Bickson, and Carlos Guestrin. PowerGraph: Distributed graph-parallel computation on natural graphs. In *OSDI*, pages 17–30, 2012.
5. Jim Gray, Adam Bosworth, Andrew Layman, and Hamid Pirahesh. Data cube: A relational aggregation operator generalizing group-by, cross-tab, and sub-total. In *ICDE*, pages 152–159, 1996.
6. Xiangnan Kong, Philip S. Yu, Ying Ding, and David J. Wild. Meta path-based collective classification in heterogeneous information networks. In *CIKM*, pages 1567–1571, 2012.
7. Grzegorz Malewicz, Matthew H. Austern, Aart J.C. Bik, James C. Dehnert, Ilan Horn, Naty Leiser, and Grzegorz Czajkowski. Pregel: A system for large-scale graph processing. In *SIGMOD*, pages 135–146, 2010.

8. Arnab Nandi, Cong Yu, Philip Bohannon, and Raghu Ramakrishnan. Distributed cube materialization on holistic measures. In *ICDE*, pages 183–194, 2011.
9. Mark Newman. *Networks: An Introduction*. Oxford University Press, Inc., New York, NY, USA, 2010.
10. Maurizio Nolé and Carlo Sartiani. Processing regular path queries on Giraph. In *EDBT/ICDT*, pages 37–40, 2014.
11. Makoto Onizuka, Toshimasa Fujimori, and Hiroaki Shiokawa. Graph partitioning for distributed graph processing. *Data Science and Engineering*, 2(1):94–105, 2017.
12. T. Pitoura and P. Triantafillou. Self-join size estimation in large-scale distributed data systems. In *ICDE*, pages 764–773, 2008.
13. Marko A. Rodriguez and Joshua Shinavier. Exposing multi-relational networks to single-relational network analysis algorithms. *Journal of Informetrics*, 4(1):29–41, 2010.
14. Yingxia Shao, Lei Chen, and Bin Cui. Efficient cohesive subgraphs detection in parallel. In *Proc. of ACM SIGMOD Conference*, pages 613–624, 2014.
15. Yingxia Shao, Bin Cui, Lei Chen, Mingming Liu, and Xing Xie. An efficient similarity search framework for SimRank over large dynamic graphs. *Proc. VLDB Endow.*, 8(8):838–849, April 2015.
16. Yingxia Shao, Bin Cui, Lei Chen, Lin Ma, Junjie Yao, and Ning Xu. Parallel subgraph listing in a large-scale graph. In *Proc. of ACM SIGMOD Conference*, pages 625–636, 2014.
17. Yingxia Shao, Bin Cui, and Lin Ma. Page: A partition aware engine for parallel graph computation. *TKDE*, 27(2):518–530, Feb 2015.
18. Yizhou Sun and Jiawei Han. *Mining Heterogeneous Information Networks: Principles and Methodologies*. Morgan & Claypool Publishers, 2012.
19. Yizhou Sun, Jiawei Han, Xifeng Yan, Philip S. Yu, and Tianyi Wu. PathSim: Meta path-based top-k similarity search in heterogeneous information networks. In *VLDB*, pages 992–1003, 2011.
20. Zhengkui Wang, Qi Fan, Huiju Wang, Kian-Lee Tan, D. Agrawal, and A. El Abbadi. Pagrol: Parallel graph OLAP over large-scale attributed graphs. In *ICDE*, pages 496–507, 2014.
21. Zhipeng Zhang, Yingxia Shao, Bin Cui, and Ce Zhang. An experimental evaluation of SimRank-based similarity search algorithms. *Proc. VLDB Endow.*, 10(5):601–612, January 2017.
22. Peixiang Zhao, Xiaolei Li, Dong Xin, and Jiawei Han. Graph cube: On warehousing and OLAP multidimensional networks. In *SIGMOD*, pages 853–864, 2011.

Chapter 6
Efficient Parallel Cohesive Subgraph Detection

Abstract Community detection is a fundamental graph analytic task. However, due to the high computation complexity, many community detection algorithms cannot handle large graphs. In this chapter, we investigate a special community detection problem, that is, cohesive subgraph detection. Here the target cohesive subgraph is k-truss, which is motivated by a natural observation of social cohesion. We propose a novel parallel and efficient truss detection algorithm, called PeTa. PeTa produces a triangle complete subgraph (*TC-subgraph*) for every computing node. Based on the TC-subgraphs, it can detect the local k-truss in parallel within a few iterations. We theoretically prove, within this new paradigm, the communication cost of PeTa is bounded by three times of the number of triangles, the total computation complexity of PeTa is the same order as the best known serial algorithm, and the number of iterations for a given partition scheme is minimized as well. Furthermore, we present a subgraph-oriented model to efficiently express PeTa in parallel graph computing systems. The results of comprehensive experiments demonstrate, compared with the existing solutions, PeTa saves $2\times$ to $19\times$ in communication cost, reduces 80% to 95% number of iterations, and improves the overall performance by 80% across various real-world graphs.

6.1 Introduction

Community structure generally exists in real-life graphs [10, 23], where nodes are more densely connected in the same community. Community detection, which aims to discover the community structure in graphs, is a fundamental problem in analyzing complex graphs [4, 8, 12, 16, 19, 22]. Cohesive subgraph [30] is one type of community structures. Most of the existing definitions of cohesive subgraph (e.g., clique [17], n-clique [3], n-clan [20], n-club [20], k-plex [27]) are faced with enormous enumeration problem (too many results) and computational intractability. For example, detecting cliques requires exponential number of enumerations since there are at most $3^{n/3}$ maximal cliques [21, 28], where n is the number of vertices in a graph.

© Springer Nature Singapore Pte Ltd. 2020
Y. Shao et al., *Large-scale Graph Analysis: System, Algorithm and Optimization*,
Big Data Management, https://doi.org/10.1007/978-981-15-3928-2_6

In this chapter, we concentrate on a new type of the cohesive subgraph, called k-truss [5], which is first introduced for social network analysis. Formally, each edge in a k-truss has at least $k-2$ triangles. We define the *support* of an edge in k-truss as the number of involved triangles. In respect of the social network, a k-truss ensures that each pair of friends has at least $k-2$ common friends, and this is consistent with sociological researches [11, 31]. Another advantage of k-truss is that it avoids the enumeration problem and can be detected in polynomial time. The complete definition of k-truss and its important properties will be described in Sect. 6.2.

However, most of the existing approaches for k-truss detection are sequential and executed on a single node [5, 29, 32]. When processing large-scale graphs, the parallel solutions are needed. Cohen [6] first proposed a MapReduce-based algorithm for the k-truss detection. As the MapReduce [7] framework is disk-oriented and unaware of graph structures, Cohen's solution is low efficiency with regard to large graphs [29]. L. Quick [25] improved the efficiency by implementing the detection algorithm on Pregel-like graph computing systems [1, 18, 26].

The above two parallel solutions [6, 25] both follow the same algorithm to detect a k truss. More concretely, they iteratively remove invalid edges in the graph according to the definition of k-truss via three sub-routines—(1) enumerating all the triangles in the graph, (2) for each edge counting the number of triangles containing it, (3) removing the edges with insufficient support. They find a valid k-truss until no more edges can be removed. Unfortunately, such algorithm still suffers from expensive communication cost and large number of iterations. Furthermore, on the basis of our theoretical analysis, the total computation complexity of the above algorithm is hardly to be close to the best known order, $O(m^{\frac{3}{2}})$, of sequential algorithms.

The root cause to the inefficiency of the state-of-the-art parallel solutions is that they conduct the repeated triangle counting and utilize improper programming models. The repeated triangle counting not only brings more iterations, but also increases the communication and computation cost. From the programming model aspect, the MapReduce is too loose to capture the graph structure, and the vertex-centric model is too restricted to operate the local graph flexibly.

To efficiently detect trusses in a large graph, we propose a new parallel truss detection algorithm, named PeTa. In PeTa, the input graph is partitioned among computing nodes, then it constructs a triangle complete subgraph (or *TC-subgraph* for short) on top of the local partition for each computing node. On the basis of TC-subgraphs, PeTa can find the local k-trusses in parallel. We prove that the global k-truss can be obtained by simply merging these local k-trusses from TC-subgraphs. In addition, PeTa avoids the tediously recounting triangles by applying the *seamless detection* (detection without restart) between successive iterations. Thus, the worst communication cost of PeTa is bounded by a tight upper limit, and the real communication cost is only related to the size of invalid triangles. The total computation complexity can achieve the same order as the best known sequential algorithm and the number of iterations is minimized.

According to the above technical contributions, we implement the PeTa with following practical optimizations. First, through empirical studies on a set of real graphs, we summarize the power-law distribution between the frequency and edges' initial supports, named as *edge-support law*. The edge-support law ensures that PeTa can achieve a low memory space cost for graphs in the real world. Furthermore, the space cost can be reduced by applying an edge balanced partition scheme. Second, we extend the classical vertex-centric model into a subgraph-oriented model. The subgraph-oriented model treats the local subgraph (partition) as the minimal computing unit and allows users to access and update the local graph directly, thus easily utilizing the existing in-memory algorithms to enhance the local performance.

We organize the remaining of this chapter as follows. The formal problem definitions are described in Sect. 6.2. In Sect. 6.3, we introduce the existing solutions and their limitations. The novel PeTa is elaborated in Sect. 6.4, and we discuss its practical optimizations in Sect. 6.5. The experimental results are presented in Sect. 6.6.

6.2 Problem Definition

In this section, we re-formalize several core concepts related to k-truss cohesive subgraph, which are first introduced by Cohen [5], followed by the definition of k-truss detection problem. Then, we describe the parallel settings for the k-truss detection problem.

6.2.1 Preliminaries

The purpose of our work is to efficiently detect the cohesive subgraph, k-truss, in an undirected graph G in parallel. The graph G consists of a vertex set V and an edge set E. An edge $e \in E$ is undirected and joins two vertices $v, u \in V$, denoted by (u, v) or (v, u). We also use $n(= |V|)$, $m(= |E|)$ to simplify the representations of the number of vertices and the number of edges in G, respectively. The symbol $d(v)$ stands for the degree of a vertex v, which equals to $|N(v)|$. $N(v)$ contains all the vertices that are adjacent to the vertex v. Moreover, the neighbor set of a vertex set V' is the union of each single vertex's neighbors, i.e., $N(V') = \bigcup_{v \in V'} N(v)$. All the notations frequently used in this chapter are summarized in Table 6.1.

A *triangle* in a graph is a cycle of length three and denoted by T_{uvw} if its three vertices are u,v,w. The *support* of an edge based on the triangle can be stated as follows.

Table 6.1 Notations summary

Symbols	Description
$G = (V, E)$	An undirected graph
n, m	The size of vertex/edge set in graph G
(u, v)	An edge in graph G
$N(v)$	All vertices adjacent to v in graph G
$d(v)$	The degree of v, i.e., $d(v) = |N(v)|$
$N(V)$	The union of $N(v)$, $v \in V$
k	The threshold for truss detection problem
$\theta_G(e), \theta(e)$	The support of an edge e in graph G
$\Gamma(G, k), \Gamma_k$	The maximal k-truss in graph G
$\overline{\rho}$	The average number of replications of an edge
γ	The edge cut ratio of a graph partitioning
T_{vuw}	A triangle formed by vertices v, u, w
$|\triangle_G|, |\triangle|$	The number of triangles in graph G

Definition 6.1 (Support) The support of an edge $e = (u, v)$ in graph G, denoted by $\theta_G(e)$, is defined as the number of triangles that the (u, v) is involved, i.e., $\theta_G(e) = |\{T_{uvw}|(u, w), (v, w) \in E\}|$.   □

When the context is clear, we simplify the $\theta_G(e)$ into $\theta(e)$. Based on the support of an edge, we proceed to define the k-truss in a graph G.

Definition 6.2 (k-truss) Given a graph $G = (V, E)$, if a subgraph $G_s = (V_s, E_s)$ in G, satisfies that $\forall\, e = (u, v) \in E_s$, $\theta_{G_s}(e) \geq k - 2$, then the subgraph G_s is a k-truss cohesive subgraph in G.   □

Furthermore, the *maximal k-truss*, denoted by $\Gamma(G, k)$, is the one that cannot be contained by any other k-truss in G. We simply use Γ_k to represent $\Gamma(G, k)$ when the context of graph G is clear. We assume that the empty subgraph is a k-truss of graph G for arbitrate k. Thus any graph G at least has a k-truss for the arbitrate k.

The following property guarantees that there is only one Γ_k in a graph G.

Property 6.1 (Uniqueness of the Maximal k-truss) Only one maximal k-truss exists in a graph G for a fixed threshold k.

Proof Suppose only q ($q>1$) different maximal k-trusses exist in a graph G, and they are $\Gamma_{k1}, \Gamma_{k2}, \ldots, \Gamma_{kq}$. According to Definition 6.2 and the maximal property, $\bigcup_{1 \leq i \leq q} \Gamma_{ki}$ is also a maximal k-truss in graph G. This leads to the contradiction that only q different maximal k-truss exist, since we now have $q + 1$ different ones. So q can only be one.   □

Here we focus on the following problem, which is the key issue for the truss-related task [5, 29, 32].

Problem (k-truss Detection) Given a graph $G = (V, E)$ and a threshold k, finding the maximal k-truss in G.

Fig. 6.1 A toy graph G and it is partitioned into P_1, P_2, P_3. The subgraph with black thick edges is Γ_4

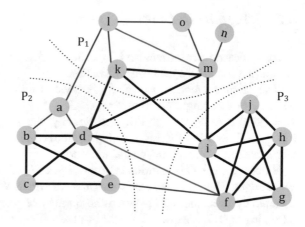

Here is an example in Fig. 6.1. Given a graph G and threshold $k=4$, the goal of above problem is to discover the subgraph with black thick edges, which is a Γ_4.

6.2.2 Fundamental Operation

The k-truss detection problem can be solved in polynomial time, but the various implementations of the fundamental operation result into different computation complexity. The *fundamental operation* is responsible for counting the triangles around a certain edge. According to Definition 6.1, $\theta(e)$ is the exact number of triangles that the edge e is involved. Therefore, $\theta(e)$ can be computed simply by counting the triangles around that edge e. Two popular solutions for the fundamental operation are listed below.

Classic Solution It first sorts the neighbors of each vertex in ascending order on their IDs. Then for an edge (u, v), the algorithm can calculate the number of triangles around (u, v) in $O(d(v)+d(u))$. This method helps discover the k-truss in $O(\sum_{(u,v)\in E}(d(v)+d(u))) = O(\sum_{v\in V} d^2(v))$ time complexity and has been applied in [5, 32].

Index-Based Solution It is the most recent solution, proposed in [29]. When operating an edge (u, v), the algorithm only enumerates neighbors w of the vertex u with smaller degree, and tests the existence of the third edge (w, v) in graph G. Since the testing can be done in constant time with the help of a Hashtable, the fundamental operation can be completed in $O(\min\{d(v), d(u)\})$. Thus, the total computing complexity is decreased to $O(m^{\frac{3}{2}})$, which is the best known time complexity.

6.2.3 Parallel Computing Context

Before proceeding to the parallel k-truss detection algorithms, we clarify the parallel computing context for the k-truss detection.

The original graph $G = (V, E)$ is divided into W partitions, which are P_1, $P_2, \ldots, P_W$. A single partition $P_i=(V_i, E_i)$, $1 \leq i \leq W$, satisfies $V_i \subseteq V$ and $E_i=\{(u, v)|u \in V_i \vee v \in V_i\}$. Additionally, the W partitions satisfy that $V = \bigcup_{1 \leq i \leq W} V_i$, and $V_i \cap V_j = \phi, i \neq j$. Initially, the W partitions are distributively stored among W computing nodes' memory.

Since the graph G is partitioned, some edges will cross different partitions, called *cross edges*. The remained edges are *internal edges*. We define the *edge cut ratio*, denoted by γ, as the ratio between the number of cross edges and the number of edges in graph G. Meanwhile, the cross edges will be replicated in order to maintain the connectivity. We use *edge replicating factor*, $\overline{\rho}$, to represent the average number that an edge is replicated. $\overline{\rho}$ is the ratio between the number of edges maintained by the parallel algorithm and the original size of E in graph G.

In the parallel computing context, all triangles in the partitioned graph G are classified into three different types according to the effect of cross edges.

- **Type I** is the triangle only consisting of internal edges.
- **Type II** is the triangle including both internal edges and cross edges.
- **Type III** is the triangle only containing cross edges.

For example, in Fig. 6.1, edges (k, m), (l, k), (l, m) are internal edges, so T_{kml} is Type I. Edges (d, k), (d, m), (d, i), (k, i) are cross edges, so T_{dkm} is Type II and T_{dki} belongs to Type III.

6.3 The Existing Parallel Algorithms

As briefly mentioned in Sect. 3.1, two parallel algorithms for k-truss detection have been developed. One is implemented on the MapReduce framework, and the other is improved by adapting the MapReduce solution to the Pregel-like systems. However, both approaches follow the same logical framework, which is first introduced by Cohen [6] and listed below.

1. Enumerate triangles.
2. For each edge, record the number of triangles, $\theta(e)$, containing that edge.
3. Keep only edges with $\theta(e) \geq k - 2$.
4. If step 3 dropped any edges, return to step 1.
5. The remaining graph is the maximal k-truss.

This type of algorithms requires the triangle counting routine in every *logical iteration* (steps 1–4). Furthermore, due to the algorithm's restriction, it will incur many iterations. Since the MapReduce framework flushes all the intermediate

---

Algorithm 1 Improved vertex algorithm for k-truss detection

---

1. /* v represents the vertex where this algorithm runs. */
2. $inerStep \leftarrow$ getSuperstep() % 3
3. **switch** ($inerStep$)
4. **case 0:**
5.   **foreach** pair $(u, w), u, w \in N(v)$ **do**
6.   send message (v, u, w) to the vertex u for completing
           edge (u, w). $\{d(v) \le d(u) \le d(w).\}$
7.   **end foreach**
8. **case 1:**
9.   **foreach** incoming message (u, v, w) **do**
10.   **if** $w \in N(v)$ **then**
11.   send message (u, v, w) to u, v. {notify the sources of
                three edges in the triangle T_{uvw}.}
12.   **end if**$\{v \in N(u) \wedge w \in N(u)$ holds from case 0.$\}$
13.   **end foreach**
14. **case 2:**
15.   **foreach** incoming message (v', u', w') **do**
16.   **if** $v = v'$ **then**
17.   increase $\theta((v, u')), \theta((v, w'))$
18.   **else if** $v = u'$ **then**
19.   increase $\theta(v, w')$
20.   **end if**$\{v = w'$ is impossible!$\}$
21.   **end foreach**
22.   **foreach** $u \in N(v)$ **do**
23.   **if** $\theta((u, v)) < k - 2$ **then**
24.   remove edge (v, u).
25.   **end if**
26.   **end foreach**
27. **end switch**

---

results to the disk and is unaware of the graph structure, it is inefficient for the iterative graph workloads [18]. The inefficiency of the MapReduce solution for k-truss detection is also pointed out in [29]. Although, the L. Quick's solution tries to eliminate the shortages of the MapReduce solution by applying the Pregel-like systems, which is in-memory and graph aware, the performance is still hindered by the internal limitations of the logical framework.

In the following subsections, we introduce our improvement on the L. Quick's solution first, and then show the inherent limitations of the existing parallel solutions.

6.3.1 Improved L. Quick's Algorithm

L. Quick's [25] parallel solution uses the popular graph computing system, Pregel, and runs on an ordered directed graph, where each edge joins two vertices from the one with small degree to the one with large degree and breaks ties based on vertices'

ID. The approach splits a logical iteration into four separated supersteps, between which a synchronization will happen. Thus, the synchronization explodes in four times. We improve it by simply getting rid of the fourth superstep, which removes the vertex with zero degree in every logical iteration. The improved approach drops the isolated vertices altogether in the end, since the isolated vertices do not bring additional computation in the middle of the algorithm. This modification not only reduces the number of synchronization (supersteps) in Pregel-like systems, but also decreases the communication cost (the original version requires messages for the vertex deletion).

The new vertex algorithm for the improved solution is listed in Algorithm 1. The algorithm still runs on an ordered directed graph, but repeats in every three supersteps until no edges are dropped. In the first superstep (Lines 4–7), each vertex v enumerates vertex pairs from its neighbors and sends corresponding triad messages for completing the third edge. The second superstep (Lines 8–13) is responsible for verifying the existence of the third edge in a triad message and notifying the sources of three edges if a triangle is found. In the third superstep, each vertex counts the number of triangles for its adjacent edges (Lines 15–21) and removes the edge whose $\theta(e)$ is below the threshold $k - 2$ (Lines 22–26).

6.3.2 Limitations Discussion

Although the parallel graph computing systems facilitate the solution and solve the problem in-memory, the approach is still hindered by several inherent limitations. First, the algorithm still enlarges the synchronized frequency by three times. Second, the repeated triangle counting brings in massive redundant computations and communications. For example, the triangles reserved in the final found k-truss will be repeatedly computed as much as the number of iterations. Third, the restriction of the vertex-centric model in Pregel only supports to implement the classic solution for the fundamental operation. This results in the computation complexity of the algorithm in one iteration being $O(d_{\max} \sum_1^n \hat{d}^2 (v))$, where $\hat{d}(v)=|\{u|u \in N(v) \wedge d(v) \leq d(u)\}|$, and $d_{\max}=\max\{\hat{d}(v)\}$.

Example Detecting Γ_4 in graph G as shown in Fig. 6.1. The improved algorithm takes three logical iterations, which totally consumes nine supersteps. The red edges and blue edges are eliminated in the first and second logical iterations, respectively. The third iteration guarantees that the result graph does not change any more. In a single iteration, a Type II or Type III triangle will cause four remote messages, in which one for querying and three for notifying. So the algorithm sends 60 messages altogether in the three iterations. However, the algorithm only occupies a small amount of memory benefit from the ordered directed graph, and the edge replicating factor $\overline{\rho}$ is one.

To summarize, these limitations inhibit existing algorithms from obtaining a satisfactory performance. We need to redesign the k-truss detection framework, and require a more flexible programming model with graph structure preserving to implement the detection framework.

6.4   The Framework of PeTa

We propose a novel **parallel** and **efficient** **truss** detection **algorithm**, called PeTa. The basic idea in PeTa is that each computing node simultaneously finds the local maximal k-truss independently. Once all the local maximal k-truss subgraphs are discovered, the computing nodes exchange the graph mutations with each other and continue to refine the previous local maximal k-truss subgraphs. The process is repeated until all the local maximal k-truss subgraphs are stable (no more external edges are removed). Finally, the maximal global k-truss subgraph is the simple union of all local maximal k-truss subgraphs.

In the following subsections, we first introduce the subgraph-oriented model, which enables parallel graph computing systems to express more rich graph algorithms, like PeTa. Then a special subgraph, TC-subgraph, is presented. Finally, the local subgraph algorithm on top of TC-subgraph in PeTa is described.

6.4.1   Subgraph-Oriented Model

The classic vertex-centric model is too constrained to (efficiently) express a rich set of graph algorithms, since a vertex algorithm is limited to access its neighbors only and the other parts of local subgraph cannot be accessed. We extend it into a subgraph-oriented model.

The subgraph-oriented model removes the vertex constraints, treats the local subgraph as the minimal operable unit, and allows users to directly access and update the local graph. Figure 6.2 shows the typical APIs for a subgraph program. For instance, at any time, user can randomly access a vertex via getVertex() method,

Fig. 6.2 API abstraction in subgraph-oriented model

```
class Subgraph{
    public abstract void subgraphCompute();
    public Vertex getVertex(long vid);
    public Vertex removeVertex(long vid);
    public void addVertex(long vid, Vertex v);
    public Iterator<Vertex> getVertexIterator();
    public Edge getEdge(long src, long dest);
    public Edge removeEdge(long src, long dest);
    public void addEdge(Edge e);
    public Iterator<Edge> getEdgeIterator();
    public void sendMessage(long vid, Message msg);
}
```

and then operate the vertex as what can be done in the vertex-centric model. The parallel computing framework executes many subgraph programs concurrently, and each of them is the minimal computing unit. The subgraph programs exchange messages with each other between the successive iterations.

Comparing to the vertex-centric model, the subgraph-oriented model is a coarse-grained one and opens up new opportunities to efficiently express local graph algorithms. Because of the flexibility of operating local graph in subgraph-oriented model, the algorithms can access vertices and edges on demand. Like in PeTa, the edges of local subgraph are visited in the removing order. It is impossible to implement such an edge access order in vertex-centric model. Furthermore, the vertex-centric model is just a special case of the subgraph-oriented model (i.e., the subgraph-oriented model becomes the vertex-centric one when the sub-graphCompute() method is implemented as sequentially visiting the vertices of local subgraph), so the subgraph-oriented model is able to express all the graph algorithms which can be realized on vertex-centric model. These graph algorithms can be optimized further by their corresponding sequential algorithms. For instance, implementing BFS on subgraph-oriented model, in each iteration, a computing node is initialized by external vertices, which have been visited by other computing nodes, and executes the sequential BFS algorithm on local subgraph. Thus the number of iterations will be reduced, since more than one-hop vertices are explored in each iteration.

To summarize the above discussion, subgraph-oriented model can express more complicated graph algorithms by eliminating the vertex constraints. It is a cornerstone for the PeTa to be expressible in a parallel graph computing system.

6.4.2   Triangle Complete Subgraph

Naturally, a raw local partition on each computing node does not include the complete information of every edge's common neighborhood, so that the Type II and Type III triangles cannot be directly captured in the local partition. Therefore, it is impossible to compute the local maximal k-truss subgraph just relying on the local partition. For instance, in Fig. 6.1, the local partition P_2 cannot be aware of the existence of triangles T_{dim} and T_{dif}, because P_2 is unable to access vertices m, i and f's neighborhoods at local.

In the PeTa, it works on the *triangle complete subgraph* to do the local computation. The definition is given as follows:

Definition (Triangle Complete Subgraph) Given a graph $G = (V, E)$, and its subgraph $G_s = (V_s, E_s)$, $V_s \subseteq V$, $E_s \subseteq E$. The triangle complete subgraph, **TC-subgraph** for short, $G_s^+ = (V_s^+, E_s^+)$ satisfies $V_s^+ = \{V_s \cup N(V_s)\}$ and $E_s^+ = E \cap$ $(\{[(u, v)| u \in V_s \vee v \in V_s\} \cup \{(u, v)|u, v \in N(V_s) \wedge \exists w \in V_s, \text{s.t.}, (w, u), (w, v) \in E_s^+\})$.   □

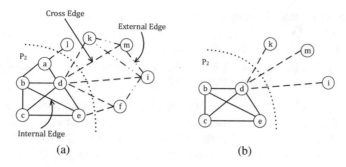

Fig. 6.3 TC-subgraph and local Γ_4 in P_2 of graph G. (**a**) TC-subgraph. (**b**) Local Γ_4

In accordance with above definition, the edges in a TC-subgraph belong to three categories. Besides the internal edge and cross edge, the third type is *external edge*. The external edge is the one whose both end vertices belong to $N(V_s)$. It is a phantom of internal or cross edge in some other TC-subgraphs. Figure 6.3a shows a TC-subgraph based on the local partition P_2 in graph G. In the TC-subgraph, (b, e) is an internal edge, (d, m) is a cross edge, and (m, i) is an external edge.

Note that the notation of TC-subgraph is more restricted than the induced subgraph. In Fig. 6.3(a), the edge (l, m) is not included in the TC-subgraph while it will be in the corresponding induced subgraph. Fortunately, Theorem 6.1 points out that the TC-subgraph has the full triangle information for all internal and cross edges, so that the TC-subgraph consumes less memory while preserving sufficient knowledge.

Theorem 6.1 *TC-subgraph $G_s^+ = (V_s^+, E_s^+)$ of the subgraph $G_s = (V_s, E_s)$ in graph $G = (V, E)$ contains the same number of triangles for every edge $e = (u, v)$, in which $(u \in V_s \vee v \in V_s)$ holds, as the graph G does.*

Proof Using proof by contradiction. Let the edge $e_0 = (u_0, v_0)$, $u_0 \in V_s$, has different number of triangles between G_s^+ and G.

(1): Assume e_0 has more triangles in G_s^+. Then there exists at least one vertex w_0, which forms a triangle $T_{u_0 w_0 v_0}$ in G_s^+ and does not form one in G. This fails to hold the condition $(u_0, w_0) \in E_s^+ \wedge (u_0, w_0) \in E$.
(2): When e_0 has fewer triangles, G contains a vertex w_0 in contrast. If w_0 is not in V_s^+, then the condition $V_s^+ = \{V_s \cup N(V_s)\}$ fails, because u_0 is in V_s, and w_0 should be in $N(V_s)$. If w_0 exists in G_s^+, then the edge (u_0, w_0), (v_0, w_0) should be included by the definition of E_s^+.

In conclusion, the theorem is correct. $\qquad\square$

Based on Theorem 6.1, the node can compute on the internal edges and cross edges locally and correctly. For simplicity, we call both edges as the *core edges* in a TC-subgraph. We proceed to define the *local maximal k-truss* in a TC-subgraph.

Definition 6.3 (Local Maximal k-truss) A subgraph $G_k = (V_k, E_k)$ in TC-subgraph $G_s^+ = (V_s^+, E_s^+)$, $V_k \subseteq V_s^+$, $E_k = E_s^+ \cap \{(u, v) | u \in (V_s \cap V_k) \vee v \in (V_s \cap V_k)\}$, is the local k-truss if and only if $\theta_{G_k'}(e) \geq k - 2$, $e \in E_k$, where $G_k' = (V_k, E_s^+ \cap \{(u, v) | u \in V_k \wedge v \in V_k\})$. The local maximal k-truss is the one which is not contained by any other local k-truss.   □

Figure 6.3b illustrates the corresponding local maximal 4-truss in the TC-subgraph of P_2. Similar to the maximal k-truss, the local maximal k-truss also has Property 6.1. The following theorem ensures the correctness of PeTa. The algorithm can discover the correct maximal k-truss with iteratively finding the local maximal ones.

Theorem 6.2 *When all the local maximal k-trusses are stable, the union of these maximal local k-trusses is the global maximal k-truss subgraph.*

Proof According to Theorem 6.1 and Definition 6.3, the local Γ_k indicates that its core edges belong to the global Γ_k if the external edges still exist in global. Since all the local Γ_ks are stable, which implies no graph mutations, the external edges in one local Γ_k must be reserved in some other local Γ_k as the core edges. So the above theorem holds.   □

6.4.3   The Local Subgraph Algorithm in PeTa

Here we elaborate the local algorithm of PeTa, which is also a subgraph program in the subgraph-oriented model. Thanks to the TC-subgraph, each computing node can detect the local maximal k-truss independently during an iteration. So the local subgraph algorithm is divided into two distinct phases. The first one is the *initialization phase*, in which TC-subgraphs are constructed and the initial $\theta(e)$ are calculated for the core edges. The other one is responsible to find the local maximal k-truss and is called *detection phase*. The pseudo-code is illustrated in Algorithm 2.

In the *initialization phase*, computing nodes generate their TC-subgraph through the triangle counting routine. The only difference is that the computing nodes should materialize triangles in local if they are Type II or Type III which involve an external edge (Line 14). Along with the TC-subgraph construction, the initial $\theta(e)$ of core edges are calculated as well (Line 12). The initialization phase takes three iterations and the third iteration mingles with the first detection iteration (Lines 9–23).

In the *detection phase*, the local maximal property ensures that each computing node can do the detection continuously between iterations without any redundant computation. Thus the whole computation on a TC-subgraph, although separated by synchronization (Line 33), is same as the one in a single node, we name it as *seamless detection*. The seamless detection starts to refine the local maximal k-truss only based on the removed external edges (Line 26), except the first detection iteration. Since the local maximal k-truss has been discovered in previous iteration, the current iteration does not need to modify it unless some external edges are

---

Algorithm 2 Local Algorithm in PeTa

Input: local partition $P = (V_i, E_i)$, threshold k.
Output: stable local maximal k-truss.
1. /*initialization phase*/
2. $G_s^+ \leftarrow P$
3. send triad message (v, u, w) to the subgraph of containing vertex u. $\{v \in V_i, u, w \in N(v), d(v) \leq d(u) \leq d(w).\}$
4. *synchronized barrier* {one iteration}
5. **foreach** incoming message (v, u, w) **do**
6.   notify v, u, w by message (v, u, w), if the edge (u, w) exists
7. **end foreach**
8. *synchronized barrier* {one iteration}
9. **foreach** incoming message (v, u, w) **do**
10.   **foreach** edge e in $\{(u, v),(v, w),(u, w)\}$ **do**
11.   **if** e is the core edge **then**
12. $\theta(e) \leftarrow \theta(e) + 1$ /*calculate initial $\theta(e)$.*/
13.   **else**
14. $E_s^+ \leftarrow E_s^+ \cup \{e\}$ /*materialize external edge.*/
15.   **end if**
16.   **end foreach**
17. **end foreach**
18. $invalidQueue \leftarrow \phi$ /* stores removable edges */
19. **foreach** core edge e in E_s^+ **do**
20.   **if** $\theta(e) \leq k - 2$ **then**
21.   enqueue e into $invalidQueue$
22.   **end if**
23. **end foreach**
24. /*detection phase.*/
25. **repeat**
26.   enqueue removed external edge into $invalidQueue$
27.   /*detect local maximal k-truss in G_s^+*/
28.   **while** $invalidQueue \neq \phi$ **do**
29. $e \leftarrow$ dequeue $invalidQueue$
30.   call Index-based solution of fundamental operation on e to decrease the support of other two edges
31.   enqueue the invalid core edges into $invalidQueue$.
32.   **end while**
33.   notify graph mutations to remote subgraphs if necessary.
34.   *synchronized barrier* {one iteration}
35. **until** All local maximal k-trusses are stable.

---

deleted in some other TC-subgraphs. Thus, $\theta(e)$ is decreased only due to the triangle missing that is related to the external edges' removal. The algorithm does not need to scan over all the local core edges in each detection iteration, only follows the edge removing order to update the local maximal k-truss. For the first detection iteration, the algorithm simply scans over all core edges once, and enqueues the edges with insufficient $\theta(e)$ to start the detection (Lines 19–23). After the local maximal k-truss is discovered, the algorithm notifies other TC-subgraphs that are influenced by the local core edges' removal (Line 32). The detection phase is accomplished when no external edges are removed.

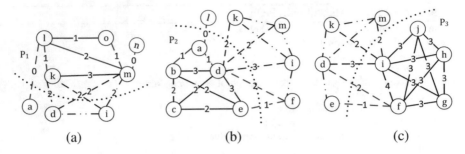

Fig. 6.4 The 3th iteration of 4-truss detection on graph G. The initial $\theta(e)$s are assigned on the core edges. (**a**) No message. (**b**) Send (e, f) removal message to P_3. (**c**) No message

Example PeTa takes four iterations to solve the 4-truss detection on the graph G shown in Fig. 6.1. The first two iterations construct the TC-subgraph. The third one initializes $\theta(e)$ of the core edges and finds the local Γ_4 successively. Figure 6.4 illustrates the third iteration. All red edges will be removed by each computing node, and only one message that causes T_{def} in P_3 missing will be sent from P_2. The last iteration refines the local Γ_4 according to the message of external edges removal and discovers all the local Γ_4s are stable. In summary, the algorithm takes four iterations and sends 25 messages, in which 24 messages are from the TC-subgraph construction.

6.4.4   Complexity Analysis of PeTa

In this section, we analyze the efficiency of PeTa. The analysis comes from four different aspects. They are space cost, computation cost, communication cost, and the number of iterations. Overall, within the new paradigm, i.e., TC-subgraph and seamless detection, the communication cost of PeTa is bounded by $3|\Delta|$. The total computation complexity is $O(\frac{\bar{\rho}^{\frac{3}{2}}}{\sqrt{W}}m^{\frac{3}{2}})$ and PeTa also achieves the minimal number of iterations for a given partition.

6.4.4.1   Space Cost

Since the TC-subgraph requires to materialize the external edges for the triangle completeness, the memory overhead is incurred, which can be measured by $\bar{\rho}$. Theorem 6.3 shows the expectation of $\bar{\rho}$ for a random partitioned graph G.

Theorem 6.3 *Given a graph G, which is random partitioned into W parts, then*

$$E\left(\bar{\rho}\right) = \left(2 - \frac{1}{W}\right) + \frac{3|\Delta|}{m}\left(1 - \frac{1}{W}\right)^2.$$

Proof In a random partition, an edge is a cross one with probability $(1 - \frac{1}{W})$, so $(1 - \frac{1}{W})m$ edges are replicated for maintaining the graph connectivity.

The other edge replication is from external edges for the completeness of Type II and Type III triangles. The number of external edges can be estimated as follows:

$$\sum_{v \in V} \left\{ \left(1 - \frac{1}{W}\right)^2 \frac{1}{2} \times \sum_{u \in N(v)} \theta((v, u)) \right\} = 3|\triangle| \left(1 - \frac{1}{W}\right)^2,$$

where $(1 - \frac{1}{W})^2$ means a triangle with two cross edges brings in an external edge.

In summary, $E(\overline{\rho})$ is

$$E(\overline{\rho}) = \frac{1}{m} \left(m + \left(1 - \frac{1}{W}\right) m + 3|\triangle| \left(1 - \frac{1}{W}\right)^2 \right)$$

$$= \left(2 - \frac{1}{W}\right) + \frac{3|\triangle|}{m} \left(1 - \frac{1}{W}\right)^2.$$

$\square$

Following corollary is directly from above theorem,

Corollary 6.1 *Given a graph G, the upper bound of $\overline{\rho}$ is $2 + \frac{3|\triangle|}{m}$.*

The upper bound in Corollary 6.1 can be achieved, when each single vertex in graph G forms a partition. When the upper bound is reached if the graph G is a clique, then every TC-subgraph becomes the entire clique. At that time, although the core edges are small, the space cost might be too large to be unacceptable. Fortunately, for the most real-world graphs, the upper bound is relatively small, and the actual $\overline{\rho}$ hardly reaches the upper bound even if the random partition scheme is used. This will be discussed in Sect. 6.5.1.

6.4.4.2 Computation Complexity

The subgraph-oriented model supports us to implement the fundamental operation in the index-based way. Thus, the parallel algorithm is able to detect the local maximal k-truss as efficient as the best serial approach [29]. Theorem 6.4 shows that the parallel algorithm also can achieve the same complexity order as the best serial one in total. This reveals that PeTa is really efficient in computation cost.

Theorem 6.4 *The total computation complexity of the parallel algorithm is* $O(\frac{\overline{\rho}^{\frac{3}{2}}}{\sqrt{W}} m^{\frac{3}{2}})$.

Proof Here we assume the random partition scheme is used, so that the edges are distributed in a perfect balance.

For the TC-subgraph construction, the main logic is the same as triangle counting, whose computation complexity is $O(m^{\frac{3}{2}})$ [2, 15].

The seamless detection technique ensures that each edge from the three types in a TC-subgraph executes the fundamental operation at most once. Combining with the index-based implementation of fundamental operation, the computation complexity in local can be $O(m_i^{\frac{3}{2}})$. Then the total complexity of the parallel algorithm is

$$\sum_1^W m_i^{\frac{3}{2}} \approx \sum_1^W \left(\frac{\overline{\rho}m}{W}\right)^{\frac{3}{2}} \quad \because \text{ Balanced Partition}$$

$$= O\left(\frac{\overline{\rho}^{\frac{3}{2}}}{\sqrt{W}}m^{\frac{3}{2}}\right).$$

(6.1)

Overall, the theorem holds when $\overline{\rho} \ll m$, which is true for the real-world graphs.

$\square$

Additionally, in accordance with the above theorem, an improvement of space cost (small $\overline{\rho}$) also enhances the total computation cost.

6.4.4.3 Communication Cost

Theorem 6.5 *The communication cost of detection phase in Algorithm 2 is bounded by* $3|\triangle|$.

Proof In detection phase, communication only occurs when the Type II and Type III triangles are eliminated. In the worst case, a triangle is removed in three independent TC-subgraphs at the same time. Moreover, Algorithm 2 guarantees that each triangle is deleted exactly once, so the communication upper bound is $3|\triangle|$. $\square$

Theorem 6.5 gives a bound of communication cost for the detection phase in PeTa. Since the real communication is only from the removals of Type II and Type III triangles, we can further decrease the communication cost with a partition which contains small number of cut triangles. This implies that a small $\overline{\rho}$ (few external edges) will reduce the communication cost as well. By the way, the communication of the TC-subgraph construction is the same with triangle counting process.

6.4.4.4 Number of Iterations

The more iterations are executed, the more synchronization is required and the heavier influence is caused by the imbalance. Therefore, it is necessary to obtain a low iteration number. Fortunately, Theorem 6.6 guarantees that PeTa achieves the minimized number of iterations for a given partition.

First, we introduce the following lemma,

Lemma 6.1 *In Algorithm 2, the number of removed edges during an iteration of the detection phase is maximized.*

Proof This can be directly derived from the algorithm. As it finds the local maximal k-truss in each iteration, there must be no core edges, whose $\theta(e) < k - 2$, in the remained TC-subgraphs. □

On the basis of Lemma 6.1, we yield

Theorem 6.6 *Given a partitioned graph* $G = (V, E)$ *and the parameter* k, *Algorithm 2 achieves the minimal iterations (synchronization).*

Proof Suppose Algorithm 2 $\mathcal{A}$ finishes in S_1 iterations, and each iteration removes e_i edges, $1 \leq i \leq S_1$. The optimal one $\mathcal{A}^*$ runs $S_2(\leq S_1)$ iterations with eliminating e_i^* in different iterations.

Property 6.1 implies that the total size of removed edges is fixed by the problem, i.e., $\sum_{i=1}^{S_1} e_i = \sum_{i=1}^{S_2} e_i^*$.

According to Lemma 6.1, the constraint $e_i \geq e_i^*$ holds. So $\sum_{i=1}^{S_2} e_i \geq \sum_{i=1}^{S_2} e_i^*$ holds as well.

Finally, we get $\sum_{i=S_2+1}^{S_1} e_i = 0$, which indicates that there are no edges removed from iteration $S_2 + 1$ to S_1. Since the termination condition of Algorithm 2 is that no graph mutation exists, so $\mathcal{A}$ must finish in S_2 iterations and $\mathcal{A} = \mathcal{A}^*$. □

6.5 The Influence of Graph Partitions

In this section, we first address the power-law distribution of initial $\theta(e)$ in real-world graphs. With the help of the skewed distribution, $\overline{\rho}$ can be small enough even a random partition is used. Then, we discuss the influences of different partition schemes on PeTa. Finally, we introduce several implementation details.

6.5.1 Edge-support Law

Through a bunch of studies on various types of real-world graphs, we find that the relationship between $\theta(e)$'s frequency and $\theta(e)$ satisfies the power-law distribution. We summarize it as the *edge-support law* as follows:

Property 6.2 (Edge-Support Law) The frequency distribution of initial $\theta(e)$ in the real-world graphs satisfies the power law, i.e.,

$$P(\theta(e)) \propto (\theta(e) + 1)^{-\alpha}, \alpha > 2.$$

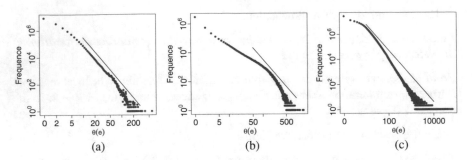

Fig. 6.5 Initial $\theta(e)$ ~Frequency distribution on various graph. (**a**) US-patent. (**b**) wWikiTalk. (**c**) DBpedia

Table 6.2 Estimated and actual $\overline{\rho}$ in real-world graphs

Graph	W	$E[\theta(e)]$	$\overline{\rho}_{est}$	$\overline{\rho}_{rand}$	$\overline{\rho}_{METIS}$
LiveJournal	32	20.00	20.74	8.99	1.77
US-patent		2.36	3.25	3.13	1.19
WikiTalk		5.93	7.53	4.52	3.31
DBpedia		7.61	9.11	6.77	2.10
LiveJournal	16	20.00	19.52	6.72	1.62
US-patent		2.36	3.14	2.95	1.15
WikiTalk		5.93	7.15	3.65	2.67
DBpedia		7.61	8.63	5.60	1.85

Figure 6.5 shows the frequency distribution of $\theta(e)$ in US-patent, WikiTalk, DBpedia as the representatives. They clearly illustrate that most of edges' support values are around zero, and only a few number of edges have a large $\theta(e)$ in the real-world graphs. Assume the edge-support law is a perfect power-law distribution, then the expectation of edge support, $E[\theta(e)]$, is

$$E[\theta(e)] = \frac{\sum_{\theta=0}^{\theta_{max}} \theta(\theta+1)^{-\alpha}}{\sum_{\theta=0}^{\theta_{max}} (\theta+1)^{-\alpha}}$$

$$\approx (\alpha-1) \int_0^{\infty} (\theta+1)^{-\alpha+1} d\theta$$

$$= \frac{\alpha-1}{\alpha-2}, \alpha > 2,$$

where $\sum_{\theta=0}^{\theta_{max}} (\theta+1)$ and $\alpha-1$ are both the normalized factor.

This implies that $E[\theta(e)]$ is a relatively small value when the perfect power law holds. Although in reality, the edge-support law is not the perfect power law, $E[\theta(e)]$ is indeed small comparing with the scale of the graph. In Table 6.2, the $E[\theta(e)]$ column shows average $\theta(e)$ across several popular graph datasets. The largest one is just 20.0 from the LiveJournal graph which contains 4.8M vertices and 43M edges.

Since $\frac{3|\triangle|}{m} = E[\theta(e)]$, on the basis of Corollary 6.1, the edge-support law ensures a small upper bound of $\overline{\rho}$. Table 6.2 also lists the estimated $\overline{\rho}_{est}$ and accurate $\overline{\rho}_{rand}$ when the graph is randomly partitioned into 16 and 32 subgraphs. We find that even if the random partition strategy is applied, $\overline{\rho}$ is small enough to be practical. For instance, the $\overline{\rho}_{rand}$ of LiveJournal can be decreased to 8.99 with 32 partitions.

6.5.2 Partition Influence on PeTa

As analyzed in Sect. 6.4.4, a small $\overline{\rho}$ reduces the space and communication cost, also improves computing efficiency. In general, we can give the formulation of computing $\overline{\rho}$ with edge cut ratio, γ, as

$$\overline{\rho} = 1 + \gamma + \gamma E[\theta(e)],$$

where the term $1 + \gamma$ represents the original edges and cut ones, while the term $\gamma E[\theta(e)]$ means whenever an edge is cut, the triangles around that edge are also cut, thus bringing external edges.

Hence, a good partition, which has small γ, can help decrease $\overline{\rho}$ further. This implies that a good partition will improve the efficiency of PeTa from space cost, communication cost, and the computation complexity. Actually, a good partition with small edge cut ratio reduces the number of iterations in heuristics as well. First, let us define θ_{min}^c (θ_{max}^c) is the minimal (maximal) $\theta(e)$ among all the cut edges. Then the number of iterations for any k belongs to the following three cases:

- $k - 2 \leq \theta_{min}^c$: No cut edges are deleted, so the algorithm finishes after the first iteration in detection phase.
- $\theta_{min}^c < k - 2 \leq \theta_{max}^c$: A portion of cut edges are deleted, and the algorithm may need other several iterations after the first iteration in detection phase.
- $\theta_{max}^c < k - 2$: All cut edges are removed during the first iteration in the detection phase. One additional iteration is sufficient to finish the whole computation.

Except the case $\theta_{min}^c < k - 2 \leq \theta_{max}^c$, the number of iterations of the other cases is deterministic, which is three and four, respectively. For the second case, a smaller $\theta_{max}^c - \theta_{min}^c$ tends to get fewer iterations. A good partition happens to meet this heuristic rule in most circumstances. Moreover, due to the local community and clustering properties [9, 24] of the real-world graphs, those graphs usually have natural groups, which results in a small γ.

However, in parallel computing, another important factor, that affecting the performance, is the balance. It is really difficult to achieve both balance and the minimal edge cuts at the same time [14]. But for PeTa, it is sufficient to obtain a partition in balance with a proper γ. For instance, simply decreasing γ from 100% to 50%, $\overline{\rho}$ can be improved at least 25%. We create the core edge balanced partition with an applicable edge cut ratio, whose implementation is described in the next subsection, to improve the performance of PeTa. Since all the computations focus on the local core edges, we balance the core edges across the partitions, not the vertices any more, for the k-truss detection. After applying the good partition scheme, we find that the six $\overline{\rho}$ of popular datasets are all below 3.5, which is shown in the last column of Table 6.2.

To sum up, the edge-support law guarantees our parallel algorithm have a small upper bound of the space cost in real-world graphs. The core edge balanced partition scheme with a small edge cut ratio improves it further in practice.

6.5.3   Implementation Details

We implement PeTa on Giraph-1.0.0 [1] by extending the vertex-centric model into the subgraph-oriented model. In the subgraph-oriented model, the logic of the subgraph algorithm follows Algorithm 2. Besides, the algorithm separately manages the core edges and external edges of a TC-subgraph. This is because the external edges only trigger the execution of subgraph program for each iteration, and during the local maximal k-truss detection, the core edges are sufficient for the computation. The fundamental operation is implemented as the index-based solution, in which the local k-truss algorithm enumerates triangles for an edge with Hashtable supported. These flexible implementations are benefit from the subgraph-oriented model. The communications between computing nodes are asynchronously executed by message passing mechanism, so the computation and communication happen concurrently.

The default partition scheme is the random partition, $\mathcal{R}$ partition for short. Furthermore, we use a good partition, which balances the core edges with a small edge cut ratio, to improve the performance of PeTa. Currently, we simply use METIS [13] to generate the reasonable partition instead of developing a new partition algorithm, and we call it as $\mathcal{M}$ partition. Since the METIS is unable to balance the core edges directly, we assign each vertex's degree as its weights, and balance the degree as an indicator for core edge balance. The degree balance factor is limited in 1%. Table 6.3 shows the γ and actual core edge balance factor, when the graphs are partitioned into 32 subgraphs by the Random and METIS methods. For a heuristic partition strategy, like $\mathcal{M}$ partition, the system requires an index to record the vertex-partition mapping schemes. Since such a map costs a small

Table 6.3 The statistics of graphs with 32 partitions

	γ		Actual balance factor	
Graph	$\mathcal{M}$	$\mathcal{R}$	$\mathcal{M}$	$\mathcal{R}$
WikiTalk	53.57%	96.87%	40%	1%
US-patent	17.93%	96.88%	1%	0%
LiveJournal	25.55%	96.89%	4%	1%
DBpedia	32.37%	96.88%	6%	0%

Table 6.4 Graph statistics

Graph	$|V|$	$|E|$	$|E|/|V|$	k_{max}
WikiTalk	2.4M	4.7M	1.95	53
US-patent	3.8M	16.5M	4.38	36
LiveJournal	4.8M	42.9M	8.84	362
DBpedia	17.2M	117.4M	6.84	52

footprint of memory, (i.e., $\sim$130M for DBpedia dataset), in current version of PeTa, each computing node stores a copy of the map to avoid communication.

6.6 Experiments

In this section, we evaluate the performance of PeTa for the k-truss detection. The experimental environment is described in the next subsection. We show the performance of PeTa can be improved by the edge balanced partition scheme, and verify the efficiency of PeTa by comparing with three other parallel solutions. At last, the scalability of PeTa is presented as well.

6.6.1 Environment Setup

All experiments are conducted on a cluster with 23 physical nodes. Each node has two 2.60GH AMD Opteron 4180 CPUs with 48 GB memory and a 10T disk RAID. Four graph datasets are used in the experiments. All the graphs have been processed into undirected ones and the meta-data of graphs are listed in Table 6.4. The DBpedia is available on KONECT,[1] and the remaining graphs are downloaded from SNAP[2] project.

Algorithms and Implementations The implementation of PeTa and $\mathcal{R}$ ($\mathcal{M}$) partition schemes used in experiments are described in Sect. 6.5.3. In addition, three other parallel solutions are compared. They are orig-LQ, impr-LQ, and Cohen-MR. **Orig-LQ** is the original L. Quick's solution [25] and **impr-LQ** are the improved one as presented in Sect. 6.3.1. Since the classic solution of counting triangle around an edge is slower than the index-based one (Sect. 6.2.2), we implement both algorithms in subgraph-oriented model, thus the index-based solution can be implemented as well, but the logical k-truss detection framework (Sect. 6.3) is reserved. **Cohen-MR** is short for Cohen's MapReduce solution and it is implemented on Hadoop-0.20.2

---

[1] http://konect.uni-koblenz.de.
[2] http://snap.stanford.edu.

using the efficient triangle counting approach [2], which counts the triangle in a single MapReduce round.

Measures We evaluate the algorithms in four different aspects, which are *space cost, communication cost, number of iterations*, and *the overall performance*. Since the space cost is only related to partition schemes and is measured by edge replicating factor, $\overline{\rho}$, which has been presented in Sect. 6.5.1, we do not illustrate it again in the following sections and readers can refer the metric $\overline{\rho}$ in Table 6.2. The smaller $\overline{\rho}$ implies a better space cost. Additionally, the value of $\overline{\rho}$ is one for all baselines. Thus, we can see that PeTa has small space overhead when $\mathcal{M}$ partition method is used.

Threshold Selection Strategy For the k-truss detection problem, a regular input parameter besides the graph is the threshold k. We denoted k_{max} as the maximal k that the k-truss of the graph is non-empty. Table 6.4 also shows the k_{max} of the four graphs. With the increase of k, the size of k-truss in a graph is shrinking, and the more edges will be deleted. Consequently, the measured metrics vary as well. In the following experiments, we only visualize the corresponding costs with several selected k. The strategy of selecting k satisfies that the sizes of results are almost uniformly distributed across all possible sizes. For instance, in Fig. 6.6c, we select ten different k (x-axis) for LiveJournal, thus the percentages of vertices of the selected k-trusses change from 36.79% to 0.06%, while the percentages of edges vary from 54.72% to 1.11%. Note that we use k and Γ_k interchangeably to refer to the x-axis in the following figures.

6.6.2   The Influence of Partition Schemes for PeTa

To verify the benefit of an edge balanced partition with a reasonable edge cut ratio (Sect. 6.5.2), we evaluate the performance of PeTa on random ($\mathcal{R}$) and METIS-based edge balanced ($\mathcal{M}$) partition schemes. Since the different W has the similar enhancement of $\overline{\rho}$ (Table 6.2) which leads to similar outcomes, we only present the results of 32 partitions, whose additionally statistics are listed in Table 6.3. The experiment results demonstrate that a partition with small edge cut ratio reduces the communication cost sharply and decreases the number of iterations as well. Therefore, the partition with small γ improves the overall performance. However, the partition with small γ cannot always guarantee to achieve a better overall performance, because the imbalance of partitions may ruin the benefit brought by the small γ in the parallel computing. The detailed explanations are presented from three different metrics.

Communication Performance Here we show that the partition with small edge cut ratio can enhance the communication efficiency. In Fig. 6.6, across four different graphs, the number of sending messages in detection phase is sharply reduced when the good partition method $\mathcal{M}$ is used. For example, $\mathcal{M}$ method can reduce 85.11%

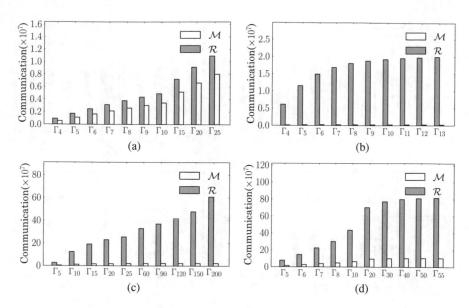

Fig. 6.6 Communication cost in detection phase. The y-axis illustrates the number of sending messages. (**a**) WikiTalk. (**b**) US-patent. (**c**) LiveJournal. (**d**) DBpedia

communications compared with $\mathcal{R}$ partition scheme to detect Γ_{10} in DBpedia. This is because more invalid edges can be removed in one iteration without extra communications on a partition with smaller γ. The improvement in Fig. 6.6(a) is not that large as others, because the close $\overline{\rho}$ between partition $\mathcal{R}$ and $\mathcal{M}$ (Table 6.2) indicates the similar cut triangles, which incurs the real communication cost. Besides, the communication tends to increase for a larger threshold k (x-axis), due to more external edges are likely to be removed as k goes up.

Number of Iterations Usually, the number of iterations can be decreased by improving the partition quality as well. In Fig. 6.7d, to detect Γ_{20} in DBpedia, the number of iterations decreases from 36 to 20 when the partition method changes from $\mathcal{R}$ partition to $\mathcal{M}$ partition. Since the good partition only heuristically (does not guarantee) reduces the number of iterations, the improvement can be small sometimes. Like detecting Γ_5 in WikiTalk in Fig. 6.7a, the improvement is zero. Another interesting observation is that the number of iterations may not increase with the increase of the threshold. For instance, it takes 10 iterations when detecting Γ_{10} in LiveJournal with good partition $\mathcal{M}$, while only 6 iterations are needed for the Γ_{60} detection, in Fig. 6.7c. This is because for a small k, some edges are deleted in several successive iterations, while those can be eliminated in one iteration when the threshold k is increased.

Overall Performance Since the overall performance is affected by several factors, such as communication, balance factor, iterations, we analyze it in three different

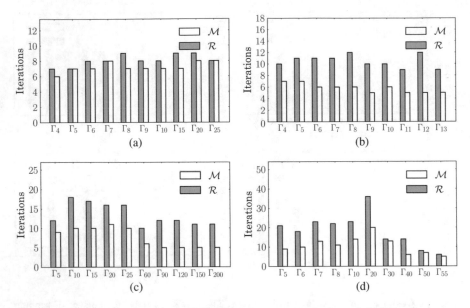

Fig. 6.7 Number of iterations. (**a**) WikiTalk. (**b**) US-patent. (**c**) LiveJournal. (**d**) DBpedia

phases. *Phase one* is the first two iterations which is responsible for exchanging the triangle information across the parallel framework. *Phase two* is the third iteration, which materializes the TC-subgraphs and executes the first detection iteration. The *last phase* completes the remaining detection.

Figure 6.8 illustrates the overall performance of k-truss detection with different partition schemes. Since phase one (bottom part) is the same with classic triangle counting on graph computing systems, it will enumerate the pairs of every vertex's neighbors. Thus, the influence of imbalance is quadratic. In Fig. 6.8, except the US-patent, all other graphs with $\mathcal{M}$ partition perform worse than the ones with $\mathcal{R}$ partition. This is caused by the imbalance of $\mathcal{M}$ partition. Refer to Table 6.3, LiveJournal, DBpedia, WikiTalk have various imbalance of core edges from 4% to 40%. Although LiveJournal only has a 4% imbalance, the performance of phase one is still decreased by around two times. However, the US-patent achieves the similar balance on both partition $\mathcal{M}$ and $\mathcal{R}$. The imbalance of partition heavily affects the performance of phase one.

The cost of phase two (middle part) is mainly determined by the number of incoming messages for materializing the TC-subgraph and the size of eliminated edges in the first detection iteration. For the fixed threshold k, except the WikiTalk, other graphs perform better with a small edge cut ratio. According to $\overline{\rho}$ in Table 6.2, these three graphs achieve a smaller space cost when the $\mathcal{M}$ partition is used, so fewer incoming messages are processed for creating external edges. However, in WikiTalk, the slight improvement of the space cost between two partition schemes reveals that the cut triangles are similar, and this leads to the similar performance

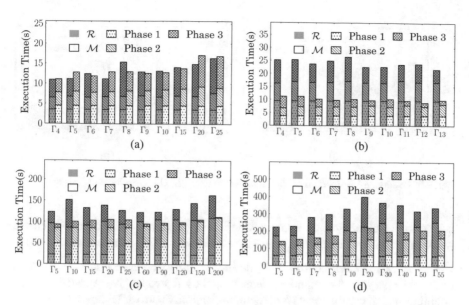

Fig. 6.8 Overall performance. From bottom to top, each bar is split into phase one, phase two, and phase three. (**a**) WikiTalk. (**b**) US-patent. (**c**) LiveJournal. (**d**) DBpedia

of the phase two. When the threshold k increases, the cost of phase two goes up as well. This is caused by more edges are deleted in the first detection iteration. A good partition which indeed decreases the space cost helps enhance the performance of phase two.

In the phase three (top part), besides a smaller ρ leads to better performance, fewer number of iterations improve the performance as well. For example, though the WikiTalk has similar $\overline{\rho}$ between two partition schemes, for Γ_8 and Γ_{15} detection, the M partition can still achieve better performance (Fig. 6.8a) in phase three by reducing the number of iterations (Fig. 6.7a).

To sum up, if a partition with small γ can reduce the $\overline{\rho}$ with affordable imbalance overhead, the overall performance can be improved as well. Otherwise, the random partition is a reasonable choice, such as the WikiTalk case, though it has a higher γ.

6.6.3 Performance Comparison

Here we compare PeTa with orig-LQ, impr-LQ, and Cohen-MR in communication cost, number of iterations, and overall performance, respectively. All the experiment results reveal that PeTa surpasses the state-of-the-art solutions in almost all aspects of the measurement even if the random partition scheme is used. Unless detecting the k-truss with large k, the existing solutions can achieve a comparable performance against PeTa. Since the results are similar on various graphs and the space

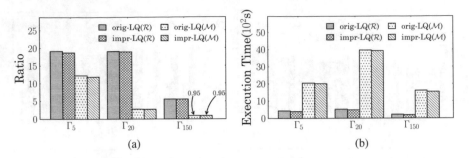

Fig. 6.9 Results on LiveJournal. (**a**) Communication ratio. (**b**) Computation comparison

is limited, the communication comparison is only presented on LiveJournal dataset, while the number of iterations comparison is shown on DBpedia dataset.

Communication Comparison PeTa detects k-truss by constructing TC-subgraphs once and applying the seamless detection, thus it avoids the repeated triangle counting, which usually incurs huge communication. Figure 6.9a shows the communication cost of orig-LQ and impr-LQ on LiveJournal with different partitions, and the cost has been normalized to the one of PeTa. It is clear to see that PeTa save the communication cost by 2X to 19X on different partition schemes. However, when k is large enough that the majority of triangles are eliminated at the first detection iteration and a few triangles are remained, the communication overhead of repeated triangle counting can be small, so that the total communication cost can be as small as or smaller than the one of PeTa. For example, when detecting Γ_{150} on LiveJournal with $\mathcal{M}$ partition, impr-LQ saves about 5% communication compared with PeTa.

Number of Iterations Comparison On all the datasets, PeTa can reduce the number of iterations around 3X to 20X on average by avoiding the repeated triangle counting and finding local maximal k-truss. Table 6.5 lists the number of iterations used for detecting Γ_5, Γ_{10}, Γ_{40} on DBpedia. For instance, when finding Γ_5, PeTa ($\mathcal{M}$) only takes nine iterations, while orig-LQ needs 2212 iterations, which is 246X higher than the former. The number in parenthesis means the count of logical iterations in the original k-truss detection framework (steps 1–4 in Sect. 6.3) of three baselines. Compared with the number of logical iterations, PeTa still achieves about 4X to 60X improvement, which is benefit from finding the local maximal k-truss. Moreover, the results reveal the existing framework is partition independent. A good

	Orig-LQ	Impr-LQ		PeTa	
k-truss	$\mathcal{R}\&\mathcal{M}$	$\mathcal{R}\&\mathcal{M}$	Cohen-MR	$\mathcal{R}$	$\mathcal{M}$
Γ_5	2212(503)	1509(503)	1006(503)	21	9
Γ_{10}	272(68)	204(68)	136(68)	23	14
Γ_{40}	112(28)	84(28)	56(28)	14	6

Table 6.5 Number of iterations on DBpedia

partition will not decrease the number of iterations, while the one in PeTa can be reduced by a good partition as analyzed in Sect. 6.6.2.

Overall Performance Comparison With the help of TC-subgraph and seamless detection, PeTa achieves a highly efficient performance on various graphs, no matter random partition or edge balanced partition is used. Figure 6.10 illustrates the overall performance of PeTa ($\mathcal{R}$), PeTa ($\mathcal{M}$), orig-LQ, and impr-LQ. Since Cohen-MR and detecting Γ_5 on DBpedia via orig-LQ and impr-LQ are at least 10X slower than the corresponding performance of PeTa ($\mathcal{M}$), their results are not visualized for figures' clarity. It is easy to figure out that PeTa performs better than all the baselines. For instance, when detecting Γ_{20} on LiveJournal, orig-LQ is 5.6X slower than PeTa ($\mathcal{M}$). The results on other three graphs are similar. Although impr-LQ performs better than orig-LQ, it still cannot win over the PeTa because of the repeated triangle counting. However, when detecting k-truss for a large k, orig-LQ and impr-LQ may achieve a comparable performance to PeTa. For instance, when detecting Γ_{40} in DBpedia, orig-LQ and impr-LQ have similar performance to PeTa ($\mathcal{M}$), and perform better than PeTa ($\mathcal{R}$). This is because, with a large k, most of triangles are eliminated early, then the overhead of repeated triangle counting is small as mentioned before.

In addition, the performance of orig-LQ and impr-LQ in Fig. 6.10 is measured on random partition, due to both baselines suffer from the good partition $\mathcal{M}$. Figure 6.9b shows the performance of orig-LQ and impr-LQ on partition $\mathcal{M}$ is 5X to

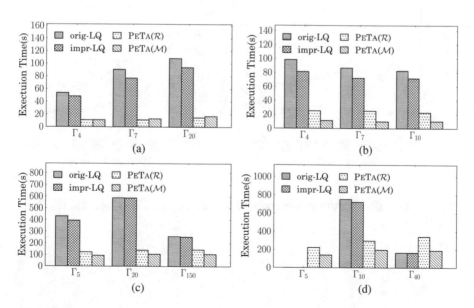

Fig. 6.10 Overall performance comparison. (**a**) WikiTalk. (**b**) US-patent. (**c**) LiveJournal. (**d**) DBpedia

8X slower than the one on partition $\mathcal{R}$. The reason is that the good partition tends to cluster the local communities in the same local partition which often belongs to the same k-truss, thus the good partition leads to a heavy imbalance of the distribution of a k-truss, while this type of imbalance seriously hinders the performance of the repeated triangle counting.

6.6.4 Scalability

Finally, to evaluate the scalability of PeTa, we run experiments of detecting Γ_{10} and Γ_{40} on DBpedia dataset with partition schemes $\mathcal{R}$ and $\mathcal{M}$. Figure 6.11a, b illustrate that, with the increase of the number of workers, the performance of PeTa improves gracefully on both partition scheme $\mathcal{R}$ and $\mathcal{M}$. For example, when detecting Γ_{10} on DBpedia with $\mathcal{M}$ partition strategy, it spends 424s with 8 workers while the cost is 184s when 32 workers are provided. However, when the number of workers increases, each partition tends to be in small size and the communication cost begins to dominate the cost for each partition, thus the improvement is not ideal linear. Furthermore, the limited number (resources) of physical computers restricts the performance improvement, so the more workers run on a single node, the slower each worker operates. When the number of workers exceeds 60 for DBpedia, the performance improvement becomes slight.

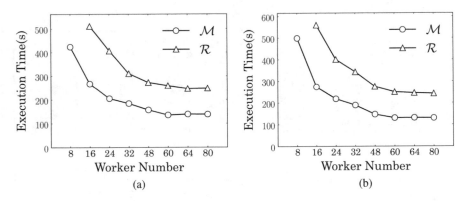

Fig. 6.11 Scalability of PeTa on different partition schemes. (**a**) Γ_{10} in DBpedia. (**b**) Γ_{40} in DBpedia

6.7 Summary

In this chapter, we developed a parallel and efficient truss detection algorithm, named PeTa. The novel algorithm has several advantages over the state-of-the-art parallel solutions, such as bounded communication cost, computation complexity with the same order as the best known serial solutions, and minimized number of iterations. Moreover, we extended the vertex-centric model into subgraph-oriented model for efficiently implementing PeTa in parallel. At last, we conducted comprehensive experiments to verify the efficiency of PeTa against the existing approaches.

References

1. Giraph. https://github.com/apache/giraph.
2. Foto N. Afrati, Dimitris Fotakis, and Jeffrey D. Ullman. Enumerating subgraph instances using map-reduce. ICDE, 2013.
3. Richard D. Alba. A graph-theoretic definition of a sociometric clique. J. Math. Sociol., 1973.
4. Zhengdao Chen, Lisha Li, and Joan Bruna. Supervised community detection with line graph neural networks. In *ICLR*, 2019.
5. Jonathan Cohen. Trusses: Cohesive subgraphs for social network analysis. NSA., 2008.
6. Jonathan Cohen. Graph twiddling in a MapReduce world. Comput. Sci. Eng., 2009.
7. Jeffrey Dean and Sanjay Ghemawat. MapReduce: simplified data processing on large clusters. OSDI, 2004.
8. Santo Fortunato. Community detection in graphs. *Physics reports*, 486(3–5):75–174, 2010.
9. M. Girvan and M. E. J. Newman. Community structure in social and biological networks. PNAS, 2002.
10. Michelle Girvan and Mark EJ Newman. Community structure in social and biological networks. *Proceedings of the national academy of sciences*, 99(12):7821–7826, 2002.
11. M. S. Granovetter. The Strength of Weak Ties. Am. J. Sociol., 1973.
12. Yuting Jia, Qinqin Zhang, Weinan Zhang, and Xinbing Wang. CommunityGAN: Community detection with generative adversarial nets. In *WWW*, pages 784–794, 2019.
13. George Karypis and Vipin Kumar. Parallel multilevel graph partitioning. IPPS, 1996.
14. Kevin Lang. Finding good nearly balanced cuts in power law graphs. Technical report, 2004.
15. Matthieu Latapy. Main-memory triangle computations for very large (sparse (power-law)) graphs. Theor. Comput. Sci., 2008.
16. Pei-Zhen Li, Ling Huang, Chang-Dong Wang, and Jian-Huang Lai. EdMot: An edge enhancement approach for motif-aware community detection. In *KDD*, pages 479–487, 2019.
17. R.Duncan Luce and Albert D. Perry. A method of matrix analysis of group structure. Psychometrika, 1949.
18. Grzegorz Malewicz, Matthew H. Austern, Aart J.C Bik, James C. Dehnert, Ilan Horn, Naty Leiser, and Grzegorz Czajkowski. Pregel: a system for large-scale graph processing. SIGMOD, 2010.
19. Pedro Mercado, Francesco Tudisco, and Matthias Hein. Spectral clustering of signed graphs via matrix power means. In *ICML*, pages 4526–4536, 2019.
20. Robert J. Mokken. Cliques, clubs and clans. Qual. Quant., 1979.
21. J.W. Moon and L. Moser. On cliques in graphs. Israel J. Math., 1965.
22. Mark EJ Newman. Detecting community structure in networks. *The European Physical Journal B*, 38(2):321–330, 2004.

23. Mark EJ Newman.   Modularity and community structure in networks. *Proceedings of the national academy of sciences*, 103(23):8577–8582, 2006.
24. M.E.J. Newman. Detecting community structure in networks. Eur. Phys. J B, 2004.
25. Louise Quick, Paul Wilkinson, and David Hardcastle. Using Pregel-like large scale graph processing frameworks for social network analysis. ASONAM, 2012.
26. Semih Salihoglu and Jennifer Widom. GPS: a graph processing system. SSDBM, 2013.
27. Stephen B. Seidman and Brian L. Foster. A graph-theoretic generalization of the clique concept. J. Math. Sociol., 1978.
28. Etsuji Tomita, Akira Tanaka, and Haruhisa Takahashi. The worst-case time complexity for generating all maximal cliques and computational experiments. Theor. Comput. Sci., 2006.
29. Jia Wang and James Cheng. Truss decomposition in massive networks. PVLDB, 2012.
30. Stanley Wasserman and Katherine Faust. Social network analysis: Methods and applications. Cambridge university press, 1994.
31. Douglas R. White and Frank Harary. The cohesiveness of blocks in social networks: Node connectivity and conditional density. Sociol. Methodol., 2001.
32. Feng Zhao and Anthony K. H. Tung. Large scale cohesive subgraphs discovery for social network visual analysis. PVLDB, 2013.

Chapter 7
Conclusions

Large-scale graph analysis is a critical task for big data applications. The distributed graph computing system is a successful paradigm for the large-scale graph analysis. It not only helps analysts achieve high scalability and efficiency, but also enables analysts to focus on the logic of analysis tasks through transparenting the tedious distributed communication protocols. In this book, we chose Pregel-like systems as a basic platform, and studied the deficiency of existing systems. We found that the systems cannot support advanced graph analysis algorithms, and do not well considered the characteristics of real-world graphs neither. We first clearly analyzed the cost of the Pregel-like systems and established the workload-aware cost model. Driven by the cost model, we further proposed several optimizations in both system level and algorithm level. In terms of the system-level optimizations, we developed a partition-aware graph computing engine PAGE. With respect to the algorithm-level optimizations, we designed three novel and efficient distributed graph algorithms for subgraph matching and cohesive subgraph detection tasks. Of course, besides the two graph mining tasks studied in this book, there still exist many other types of graph-based algorithms. For example, SimRank [2, 4] is a fundamental and useful graph statistic, and also extensively studied in serial algorithms. Due to its high computation complexity, it is almost impossible to efficiently compute the exact value. For such tasks, we need to design approximated solutions before developing distributed solutions. Overall, the current mature distributed graph computing systems are designed for a commodity cluster and focus on classical graph analysis tasks.

Next we briefly discuss two interesting directions on efficient graph analysis in the coming artificial intelligence era.

1. New Hardwares The rapid development of new hardwares, e.g., GPU, FPGA, provides us much more computation ability than before. They have been successfully used in certain domains. For example, although GPU has been extensively used in deep learning area, a general graph analysis system on top of GPU is still an ongoing work. Gunrock [3] is a relative promising project, which is a CUDA

© Springer Nature Singapore Pte Ltd. 2020
Y. Shao et al., *Large-scale Graph Analysis: System, Algorithm and Optimization*,
Big Data Management, https://doi.org/10.1007/978-981-15-3928-2_7

library for graph processing designed specifically for the GPU. It uses a high-level, bulk-synchronous, data-centric abstraction focused on operations on vertex or edge frontiers. Some works [5] also began to adopt FPGA to graph analysis, and introduce vertex-centric graph computation model for FPGA. Furthermore, graph analysis on top of heterogeneous computing resources (e.g., hybrid of CPU, GPU, and FPGA) is a much more challenging task.

2. New Applications Graph neural network, especially, graph convolutional network [1], is one of the most popular deep learning models. It has been applied in various applications, such as recommendation, text analysis, and so on. It is a kind of graph-based computing graph; however, the primitive operator is a totally different classical graph operator. The former only uses the graph structure to model the dependency between variables and the underlying computation is matrix computation. This new computation pattern makes the current distributed graph computing system be failed. Existing algorithms or libraries cannot easily scale to train large-scale graph neural networks. It is an urgent task to develop a general and scalable system for the graph neural network.

References

1. Thomas N. Kipf and Max Welling. Semi-supervised classification with graph convolutional networks. In *5th International Conference on Learning Representations, ICLR 2017, Toulon, France, April 24–26, 2017, Conference Track Proceedings*, 2017.
2. Yingxia Shao, Bin Cui, Lei Chen, Mingming Liu, and Xing Xie. An efficient similarity search framework for SimRank over large dynamic graphs. *Proc. VLDB Endow.*, 8(8):838–849, April 2015.
3. Yangzihao Wang, Andrew Davidson, Yuechao Pan, Yuduo Wu, Andy Riffel, and John D. Owens. Gunrock: A high-performance graph processing library on the gpu. In *Proceedings of the 21st ACM SIGPLAN Symposium on Principles and Practice of Parallel Programming*, PPoPP '16, pages 11:1–11:12, New York, NY, USA, 2016. ACM.
4. Zhipeng Zhang, Yingxia Shao, Bin Cui, and Ce Zhang. An experimental evaluation of SimRank-based similarity search algorithms. *Proc. VLDB Endow.*, 10(5):601–612, January 2017.
5. Shijie Zhou, Rajgopal Kannan, Hanqing Zeng, and Viktor K. Prasanna. An FPGA framework for edge-centric graph processing. In *Proceedings of the 15th ACM International Conference on Computing Frontiers*, CF '18, pages 69–77, New York, NY, USA, 2018. ACM.

Printed in the United States
by Baker & Taylor Publisher Services